Turf Grass: Pesticide Exposure Assessment and Predictive Modeling Tools

ACS SYMPOSIUM SERIES **1028**

Turf Grass: Pesticide Exposure Assessment and Predictive Modeling Tools

Mary T. Nett, Editor
Water Quality Consulting

James N. Carleton, Editor
US EPA Office of Water

Joseph H. Massey, Editor
Mississippi State University

Sponsored by the
ACS Division of Agrochemicals

American Chemical Society, Washington DC

Library of Congress Cataloging-in-Publication Data

Turf grass : pesticide exposure assessment and predictive modeling tools / Mary T. Nett, editor ; James N. Carleton, editor ; Joseph H. Massey, editor ; sponsored by the ACS Division of Agrochemicals.
p. cm. -- (ACS symposium series ; 1028)
Includes bibliographical references and index.
ISBN 978-0-8412-7421-1 (alk. paper)
1. Pesticides--Environmental aspects--Congresses. 2. Turfgrasses--Environmental aspects--Congresses. 3. Runoff--Environmental aspects--Congresses. 4. Runoff--Simulation methods--Congresses. 5. Water--Pollution--Congresses. I. Nett, Mary T., 1952- II. Carleton, James N. III. Massey, Joseph H. IV. American Chemical Society. Division of Agrochemicals.
TD427.P35T87 2009
635.9'642995042--dc22

2009044955

The paper used in this publication meets the minimum requirements of American National Standard for Information Sciences—Permanence of Paper for Printed Library Materials, ANSI Z39.48–1984.

Distributed by Oxford University Press

PRINTED IN THE UNITED STATES OF AMERICA

Foreword

The ACS Symposium Series was first published in 1974 to provide a mechanism for publishing symposia quickly in book form. The purpose of the series is to publish timely, comprehensive books developed from the ACS sponsored symposia based on current scientific research. Occasionally, books are developed from symposia sponsored by other organizations when the topic is of keen interest to the chemistry audience.

Before agreeing to publish a book, the proposed table of contents is reviewed for appropriate and comprehensive coverage and for interest to the audience. Some papers may be excluded to better focus the book; others may be added to provide comprehensiveness. When appropriate, overview or introductory chapters are added. Drafts of chapters are peer-reviewed prior to final acceptance or rejection, and manuscripts are prepared in camera-ready format.

As a rule, only original research papers and original review papers are included in the volumes. Verbatim reproductions of previous published papers are not accepted.

ACS Books Department

Contents

Indexes

Preface

The majority of Americans reside in urban or suburban settings where well-maintained turf grass provides numerous aesthetic, safety and ecological benefits. By one estimate, up to 90% of Americans may come into some contact with grass on a given day. This contact could be in the form of school children playing kickball during recess, an office worker walking to lunch with colleagues, a golfer practicing at the local driving range, a soccer team kneeling for a pre-game pep talk from their coach or a home-owner puttering about the lawn on a relaxing weekend. As current turf management practices frequently involve the use of pesticides and other crop protection chemicals, there exists a need to continuously improve our ability to assess the environmental and/or human health risks which may be posed by these compounds.

In addition to pollution and drinking water issues, there are concerns surrounding the widespread population declines observed for amphibians, insect pollinators, and endangered species such as salmon. Results from the United States Geological Survey's National Water-Quality Assessment program (USGA-NWQA) indicate that pesticides were detected in urban settings at concentrations similar to, but greater overall frequencies than, those in agricultural areas. Pesticides used on turf grass are being detected in urban surface and ground waters; the verdict is still out on the role that this category of pesticide use may play, if any, in regional environmental impacts and declines. Regardless, the need for improved procedures and models to assess potential ecological exposure from pesticides remains high.

This book, *Pesticide Exposure Assessment and Predictive Modeling Tools: Turf Grass*, is based on a symposium held at the 230th National Meeting of the American Chemical Society in Washington, DC. It presents advances made in techniques used for measuring, modeling and assessing the human and ecological effects of pesticides applied to turf grass. The book chapters are ordered in three sections: the first presents exposure assessments; the second, field studies; and the third, probabilistic modeling methodologies.

Chapter 1

Research on the Fate of Pesticides Applied to Turfgrass: A Perspective by a Scientist, Administrator and Emeritus

Al Smith[1, 2]

[1]Department of Crop and Soil Sciences, University of Georgia, Griffin, GA
[2]Current Address: 1010 Whatley's Mill Lane, Greensboro, GA

During the final two decades of the last century, there appeared to be an increasing concern for the potential movement of chemicals from intensively managed turfgrass. Funding opportunities appeared and numerous research programs were initiated across the United States. Results of a research program conducted at the University of Georgia from 1992-1997 indicated that very small fractions of certain pesticides were transported through lysimeters containing the soil mixture recommended by the United States Golf Association for turfgrass maintained as golf course greens. Additionally, small quantities of certain pesticides were transported in surface runoff from treated mini-plots simulating golf course fairways and home lawns. It was concluded that certain pesticides could be applied to turfgrass with minimal risk. Other research programs, conducted during the 1990's, obtained similar results and reached similar conclusions. Did these publicized conclusions result in apathy toward risk assessment research on turfgrass management? It appears that the importance of risk assessment studies on turfgrass management strategies has lessened during recent past. Special funding (eg. grants and gifts) has been reduced considerably over the past five years. Additionally, reduced state and federal budgets have impacted the formula funding at Land Grant Universities (LGU's). Currently in LGU's, use of formula funding for this research suffers compared to other disciplines (eg. genomics, proteomics, and metabolomics). The clientele of these Universities demand research programs for improved production and profitability. Administrators of LGU's are faced with tough decisions. The lack of funding sources, the absence of a clientele support, and the apparent apathy toward the data by regulatory agencies create a dilemma for the administrators of LGU's when it comes to utilizing their limited funding for risk assessment programs. Risk assessment/management research programs need: organizations (eg. regulatory agencies) that command their data; a clientele base that depends on the data for the profit margin; and funding agencies.

Introduction

The increasing importance of management practices utilized on golf courses has made it necessary to evaluate the environmental impact of these practices. Generally, perennial grasses have been considered to be a vegetation type that offers stability and preservation to ecosystems. The reduced cultural practices necessary to sustain a perennial crop conserve the soil, compared to the planting and maintenance of an annual crop. The extensive fibrous root system of a perennial grass system increases the soil-water infiltration rate compared to most annual herbaceous crops. Finally, year-long ground cover is usually greater for a grassed area compared to other cropping systems. The benefits of turfgrass as a ground cover, compared to other vegetation types are discussed by Smith (*1, 2*).

The maintenance of a high quality sod for use as golf course greens and fairways requires management strategies that are not always perceived as friendly to the environment. Strategies that include chemical inputs have become a major concern for the press, and ultimately the populace, and these concerns have been translated into the need to develop an acceptable data base to determine the impact of certain golf course management strategies on the environment. Currently, there are more than 16,000 golf courses in the United States. Assuming the average size of 48.6 ha per course, there are nearly 800,000 ha of turfgrass in the golf course industry receiving aggressive management strategies. Nearly 30 million U.S. golfers enjoy these courses and recognize the need for aggressive management systems.

Assuming that 2% of a golf course is managed as putting greens, there are 16,000 ha of greens in the USA that are constructed for maximum infiltration and percolation of water through the rooting media, terminating in a drainage system (e.g. drainage ditch, etc.). Fairways comprise approximately 98% of golf courses and are typically intensively managed, resulting in soil moisture content maintained near field capacity. The fairways are developed on soils typical for each region, and in the Piedmont region, these soils have a high clay content allowing for low water infiltration rates. As much as 70% of a moderate intensity rainfall will occur as overland runoff from the sloped areas typical of the Piedmont region (*3*). This water from the greens and fairways can eventually terminate in potable water containments.

Research Basis for Perspectives

A research program was developed by faculty at the University of Georgia (UGA) to determine the potential fate of pesticides applied to simulated golf course greens and fairways. The objectives of the research program were to evaluate the potential movement of pesticides and fertilizer components following application to golf courses, and to develop Best Management Practices to reduce the potential for analyte transport to potable water systems. The initial steps for evaluating the potential movement of certain pesticides were accomplished using pesticides registered for use on golf courses (Table I) on simulated greens and fairways constructed at the Griffin Campus of UGA (*3*). Simulated greens and fairways were constructed, and pesticide-analytical procedures were developed or improved (*4, 5, 6, 7*) to determine the movement of certain analytes through golf course greens and from golf course fairways.

The construction of golf course greens according to United States Golf Association specifications resulted in rapid infiltration and percolation of water through the rooting medium and out the drain system into surface drainage areas (*8*). At first inspection, these characteristics seemed to allow for the movement of large quantities of pesticides into surface drainage areas. However, our data indicated that the concentrations and quantities of pesticides transported through the simulated greens were very low (Table II). The more water soluble pesticides (eg. 2,4-D; dicamba and mecoprop) were found to have short residence time under the sod. We found that these pesticides were degraded rapidly in the moist high-organic matter media (A. Smith, unpublished). Our data indicated that the half-life for 2,4-D was less than one week at temperatures higher than 17°C (unpublished data). The pesticides with lower water solubilities (eg. dithiopyr, chlorothalonil and chlorpyrifos) had higher soil sorption capacities, increasing their residence time in the rooting medium (because of the sphagnum peat moss component) and allowing for degradation even if the half-lives were longer. This concept was best demonstrated with dithiopyr (*9, 10, 11, 12*).

Table I. Pesticides[§] and Rates Used in This Research

Common Name	*Pesticide Chemical Nomenclature*[a]	*Rate (kg/ha)*
Benefin	*N*-butyl-*N*-ethyl-2,6-dinitro-4-(trifluoromethyl) benzenamine	1.70
2,4-D DMA[b]	(2,4-dichlorophenoxy) acetic acid	2.24
Dicamba DMA	3,6-dichloro-2-methoxybenzoic acid	0.56
Dithiopyr	S,S-dimethyl 2-(difluoromethyl)-4-(2-methylpropyl) -6-(trifluoromethyl)-3,5-pyridinedicarbothioate	0.56
Chlorothalonil	2,4,5,6-tetrachloro-1,3-benzenedicarbonitrile	9.50
Chlorpyrifos	*O,O*-diethyl *O*-(e,5,6-trichloro-2-pyridyl) phosphoro-thioate	1.12
Mecoprop DMA	(±)-2-(4-chloro-2-methylphenoxy) propanoic acid	1.68
Pendimethalin	N-(1-ethylpropyl)-3,4-dimethyl-2,6-dinitrobenzen amine	1.70

[§] Transport of fertilizer-derived nitrate-N was also monitored
[a] International Union of Pure and Applied Chemistry
[b] DMA = dimethylamine salt formulation

Table II. Pesticide Transported from Field Lysimeters Under Penncross Bentgrass or Tifdwarf Bermudagrass

		Maximum Total Residue Transported Over 70 Days	
Pesticide	*Application Rate kg/ha*	*μg/L*	*% Applied±SE*
2,4-D DMA[a]	0.28	3.2	0.50±0.04
Dicamba DMA[a]	0.07	3.6	0.20±0.16
Mecoprop DMA[a]	0.56	3.8	0.20±0.14
Dithiopyr EC[b]	0.56	2.4	0.49±0.26
Dithiopyr G[c]	0.56	1.7	0.44±0.32
Chlorpyrifos	1.14 (monthly)	7.2	0.01±0.01
Chlorothalonil	9.50 (2x monthly)	2.6	0.01±0.01
OH Chlorothalonil[d]	Not Applicable	160.0	0.10

[a] Dimethylamine salt analyte
[b] Emulsifiable concentration formulation
[c] Granule formulation
[d] Metabolite of chlorothalonil from lysimeters treated with chlorothalonil

Although pesticide metabolites were not routinely analyzed, we chose to determine the transport of the more polar metabolite of chlorothalonil (hydroxychlorothalonil) in effluent from lysimeters treated with chlorothalonil. Data (Table II) indicate that concentrations as high as 160 *ug* L^{-1} were determined in the effluent from the lysimeters treated with chlorothalonil. Similar information is reported by Armbrust (*13*). This is not to imply that this is a concentration of great concern, but only to point out that first order metabolites of the pesticides should be considered in future research.

Losses of large quantities of water as surface runoff from fairways are not uncommon, and in some areas of the U.S., as much as 70% of the incoming water from an average rain event can be lost from the surface of a soil with a moisture content near the saturated condition (*14, 15, 16, 17*). Our simulated fairways were developed on a kaolinite-clay loam soil with a 5% slope. As much as 40% of the rainfall left the surface of the plots if the rain event occurred when the soil moisture content was near field capacity. Also, the simulated rainfall intensity of 3.3 cm hr^{-1}, used in our research, is not uncommon for summer rain events in the Piedmont Region of Georgia.

Analytes with the highest water solubility were found in highest concentration in water collected during the first rainfall event at 24 hr after treatment. The concentrations of nitrate-N, mecoprop, 2,4-D and dicamba, in the runoff water from this rain event, were 12,000, 810, 800, and 360 *ug* L^{-1}, respectively. The less water soluble analytes (benefin, pendimethalin, dithiopyr, chlorothalonil, and chlorpyrifos) were transported at lower concentrations.

The relationship of the analyte fraction transported to the log of the analyte water solubility (pSw) was better fit by a quadratic (R^2=0.96) than a linear function (R^2=0.86) (Figure 1). Higher fractions of water soluble analytes were transported from the treated plots over the duration of the treatment period.

The concentrations of nitrate-N in the runoff water collected 24 hours after treatment (HAT) were slightly above the recommended (USEPA guidelines) maximum contaminant levels (MCL) in potable water of 10,000 *ug* L^{-1}. The concentration of 2, 4-D, in the runoff water was above the recommended MCL of 70 *ug* L^{-1}. Although the treatment conditions were not worst-case-scenarios, there were some conditions that were near optimum for maximum runoff. The soil moisture in the treatment plots was near field capacity at the time of treatment, with a 2.5 cm rain simulation applied to the area 24 hr prior to the treatment. Rainfall in the southern Piedmont Region approximates 2.5 cm per week. At 24, 48, 96 and 192 HAT, the plots received simulated rainfall events at averages of 5.0, 5.0, 2.5, and 2.5 cm, respectively. Therefore, the total weekly simulated rain events were above the average weekly rainfall. Only samples collected over the first 192 HAT contained concentrations of the analytes capable of being detected. The average fractions of water leaving the plots as runoff following the respective simulated rain events were 44.8, 72.1, 40.0, and 35.5% of applied (3). The highest concentrations of the pesticides in the runoff water occurred during the first simulated rain event applied at 24 HAT, and approximately 84% of the recovered analytes were transported during the first two simulated rain events.

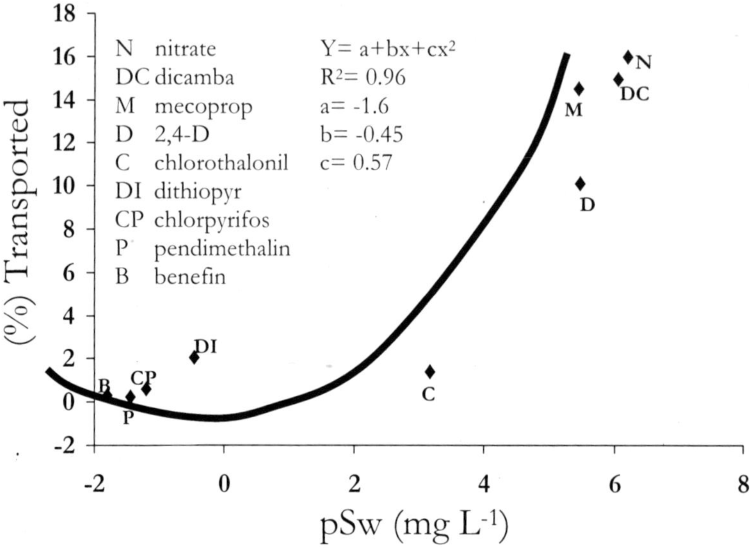

Figure 1. Fraction of the applied pesticides transported from simulated fairways as a function of the log of the analyte water solubility (pSw).

Risk Assessment, Risk Management, and Risk Reduction were phrases commonly used during the presentation of this research. Even though the nature of the risk was not identified, it was apparent that the movement of pesticides into the environment could be managed. We invoked management practices to determine if we could reduce the fraction of water soluble analytes transported from the treated sites. Unpublished research data indicated that the pesticide formulation (salt vs. ester) did not reduce the concentration or the fraction of analyte transported from the treated site at 24 HAT. A buffer area, between the terminus end of the treatment and the collection site, did not affect the fraction of analytes transported, and the concentration was only affected by the dilution factor (ie, less plot area was treated with pesticide). Soil moisture content of 10.9% (near wilting point) at the time of the first simulated rain event (24 HAT) reduced the analyte and the quantity of runoff water by 66% compared to a soil moisture of 18.5% (near field capacity) (unpublished data). Additionally, it was determined that applications of pesticides at the 10.9% soil moisture content followed by a light (1.5 cm) irrigation at 4 HAT reduced the concentration of 2,4-D in the runoff water, at the 24 HAT rain event, to 73 *ug* L^{-1}. This is a ten fold decrease compared to the treatment without the intermediate light irrigation. This would indicate that golf course superintendents could reduce the risk by applying irrigation water at periods following treatment, without reducing pesticide efficacy. Pressure injection of the pesticide at 21.3 MPa reduced the fraction of the insecticide, trichlorfon, transported over the 192 hr treatment period by 80% and the concentration in the 24 HAT collection by 95%, compared to data from the application at 166 kPa (unpublished data). Pressure injection did not increase the transport of trichlorfon through the greens media. A simple change in application technology could result in risk reduction.

Perspective of a Scientist

The age-old question seems to be "Are Golf Courses Friend or Foe of the Environment?" As a scientist, I maintain that grass has such a positive effect on the environment, compared with other crops, and that a manager would have to insult the environment with harmful management practices to negate the positive.

The Bible specifies that grass was ordained by the Creator to be the first life on Earth. "And he said let the earth bring forth grass and the earth brought forth grass…and the evening and the morning were the third day (Genesis 1:11-13)." A blade of grass is the alpha (the beginning) of the visible organic molecules. Grass takes carbon dioxide and water and manufactures complex organic molecules. If the molecules are not in its own domain, it furnishes the intermediates for the grazing animal to finish the manufacturing process. Approximately 50% of the 0.9 billion hectares of land area in the United States are covered with grass; 12 million of those hectares are managed as turf, and 0.8 million are managed as golf courses. It must be pointed out that grass preceded the golfer by several million years as he was brought forth on the sixth day. Walt Whitman wrote "I believe that a blade of grass is no less than the journey-work of the stars." The benefits of grass to the ecosystem have been summarized by Smith (1).

As good stewards of the environment, it is realized that we should continue to lessen the impact of crop management practices, even though the effects of these practices may seem miniscule. Our data indicate that some of the pesticides applied to golf courses have the potential to move into potable water systems. These data were generated from samples taken at the terminus end of the simulated fairway plots and directly under the greens media. It must also be realized that there are many fold (tens of thousands) dilutions occurring to runoff water as it moves toward potable water systems.

The critical issue facing research and regulatory institutions responsible for turfgrass management is the development and interpretation of data on the environmental fate and safety of pesticides used in the management of turfgrass on recreational facilities and home lawns. Safety cannot be measured, but risk can be estimated. Things are deemed safe if their attendant risks are judged to be acceptable. The rapid growth of the turfgrass industry during the last decade placed an urgency on the need for risk assessment of turfgrass management strategies. Risk assessment has always been with us. When cave men recognized that animals could be a source of food, they had to weigh the hazards of being mauled against starvation. There are writings about risk assessment that date back 3,000 years, yet the present concern began in 1960.

Risk management for pesticides begins by decreasing the potential dose through reducing the quantity of a compound in potable water systems. It would be desirable for there to be zero-levels of xenobiotics in potable water systems. Success in the technological development of efficient methods and ultra-sensitive instruments for detecting pesticides has resulted in the identification of some pesticides in water that would not have been detected (zero-level) several years ago.

Therefore, much of the concern for pesticides in drinking water has evolved from quantification of compounds which, because of their constituents, can now be detected at subpart per billion levels. Once the part per million was a visible limit; now we commonly measure analytes in parts per trillion. We will achieve common recognition of a part per quadrillion in the next decade. The 'zero-level' is continually pushed down and we need to recognize what is reasonable for zero-level.

Human risks from xenobiotics is generally defined as Dose x Toxicity. Presently, scientists routinely measure concentrations of pesticides in water to levels of parts per trillion. The USEPA has been working to establish drinking water standards of reference doses for chemicals in surface and ground water, based on the same toxicological research used to establish reference doses (formerly called Acceptable Daily Intake) for food. Until these or similar standards are established by USEPA, it will not be possible to assess the human water-ingestion risk from pesticides that enter the environment.

In hind sight, the following questions should be asked of research programs, such as ours, that quantified potential doses of pesticides where the toxicity is a unitless entity:

*Who really cared?
*Were our data, written in the numerous publications, utilized?
*Was there a demand for more data of this type?
*Was there a clientele for this data?
*Was there a demand for environmental fate data of another type (watershed scale)?
*Was the apparent reduction in funding for pesticide-fate research a reality?

Perspective of an Administrator

Upon entering administration as head of the Department of Crop and Soil Sciences at UGA in 1997, I found that these questions had to be answered for all research programs. The next seven years of my tenure at UGA were laced with decisions on program development, to include filling new and vacated positions for the benefit of the department. This was during a time of reduced federal and state budgets compared to the late 1980's and early 1990's, which directly impacted the formula funding available for program maintenance and development. There was strong competition for positions, and these positions had to be justified by importance to the clientele and potential for generating external funding. The previous questions had to be answered when considering continued funding for the pesticide chemistry program at the Griffin Campus UGA.

The decision for filling either a position in Crop Biotechnology which would be funded by a \$1.5M Eminent Scholar Endowment, or a Pesticide Chemistry position with \$150K start up funding was not rocket science. At the time there was no way to hire a faculty member in pesticide chemistry with assurance of adequate funding necessary to maintain the high-maintenance laboratory and the necessary technical assistance. During the early 1990's, much of our funding was obtained from the United States Golf Association (Greens Section), Golf Course Superintendents Associations, and formula funding. These funding opportunities decreased greatly at the turn of the century. Historically, the chemical industry has not funded Risk Assessment research to the level that they continue to fund pesticide efficacy research. We received very little funding from the industry for our Pesticide Fate Research Program. Equally important to the decision making was the consideration of the need for the research data. There was no apparent clientele or demand for the data.

Perspective of an Emeritus

Risk Assessment for turfgrass management systems needs to be supported by a consortium of scientists from USEPA, the turfgrass industry, and research institutions. USEPA decides the importance of data and models by defining and enforcing "acceptable-potential risk" based upon the presence of pesticides in the environment, and toxicity to ecosystem components. They will have to incorporate "acceptable levels of environmental risk" into the guidelines for registration and re-registration of pesticides.

The turfgrass industries, including chemical companies, will have to provide the pool of funding for academic research programs. Scientists from research institutions and the chemical industry will provide the unbiased research data important to decisions on risk management. In the past, funding was provided to a number of research programs without coordination of the data type to be accumulated. Field research was performed on plots of various sizes. Laboratory analyses were not unified and data quality was not monitored nor regulated by a uniform Good Laboratory Practices program for comparison of data. This must be rectified to minimize the cost of research while increasing the quality of research. It may be necessary to analyze all water samples at one location to minimize the costs for maintaining several expensive-analytical laboratories. Scientists will have to forgo the pride of maintaining individual programs.

Swan Song

As I overlook the 17th and 18th fairways and greens on the Oconee golf course at Reynolds Plantation and absorb the beauty of the water, forest, and grass environments, I am filled with pride to have been a small part of the research efforts devoted to decreasing the impacts of management practices on these beautiful components of the ecosystem.

References

1. Smith, A. E. *Through the Green*. March/April, **1993**, p 10.
2. Smith, A. E. *Grounds Maintenance*, March, **1996**, p 14.
3. Smith, A. E.; Bridges D. C. *Crop Sci.* **1996**, *36*, 1439-1445.
4. Hong, S.; Lee, A. E.; Smith, A. E. *J. Chromatog. Sci.* **1994**, *32*, 83-85.
5. Lee, A. E.; Hong, M. K.; Smith, A. E. *J.AOAC Inter.* **1994**, *77*, 1077-1083.
6. Lee, A. E.; M. K. Hong; Smith, A. E. *J. AOAC Inter.* **1995**, *78*, 1459-1464.
7. Hong, M. K.; Kim, M. C.; Smith, A. E. *J. AOAC Inter.* **1996**, *79*, 998-1004.
8. USGA Green Section Staff, *Green Section Record.* March/April **1993**, p 2.
9. Hong, S.; Smith, A. E. *J. Environ. Sci. and Health.* **2001**, *36*, 529-543.
10. Hong, S.; Smith, A. E. *J. of Agric. and Food Chem.* **1996**, *44*, 3393-3398.
11. Hong, S.; Smith, A. E. *J. Environ. Qual.* **1997**, *26*, 379-385.
12. Smith, A. E.; Bridges, D. C. *Commun. Soil Sci and Plant. Anal.* **1996**, *27*, 2381-2395.
13. Armbrust, K. L. *Pest Manag. Sci.* **2001**, *57*: 797-802.
14. Ma, Q. L.; Smith, A. E.; Hook, J. E.; Bridges, D. C. *Pest. Sci.* **1999**, 423-433.
15. Ma, Q. L.; Smith, A. E.; Hook, J. E.; Smith, R. E.; Bridges, D. C. *J. Environ. Qual.* **1999**, *28*, 1463-1473.
16. Smith, A. E. *Inter. Turfgrass J.* **1997**, 197-204.
17. Smith, A. E.; Bridges, D. C. In *Chemistry of Herbicide Metabolites in Surface and Ground Water*; Meyer, M.T.; Thurman, E. M. Eds.; ACS Symposium Series 630; Am. Chem. Soc. Press. **1996**, 165-177.

Chapter 2

Aesthetics and Practice of Maintaining the Ideal Lawn in Peachtree City, GA

Ted L. Gragson[1] and Andrew G. Keeler[2]

[1]Department of Anthropology, University of Georgia, Athens, GA 30602
[2]School of Public Policy and Management, Ohio State University, Columbus, OH 43210

In Peachtree City, Georgia, household income has a positive relationship to the money spent on pesticides, herbicides, and fertilizers and a statistically significant relationship with total expenditures on lawn and garden maintenance. Households spent 20 hours and $103 per year applying herbicides, pesticides and synthetic fertilizers, and $921 annually on lawn and garden maintenance. Peachtree City residents expended labor and money in varied ways on their lawns, although the activities with the potential for environmental damage were not universally practiced. *Who* adopts particular practices or *why* they adopt them may ultimately be less important in suburban landscapes than understanding *where* and *when* certain behaviors occur.

The popular preference for, and attachment to, the residential lawn has been well documented (*1-4*), and even the environmentally aware possess an unwavering preference for the residential lawn aesthetic (*5*). This suggests that lawns serve a purpose for many people that is more important than the ecological integrity of their surroundings.

To realize the residential lawn aesthetic requires numerous homeowner decisions about lawn care and the application of turf care chemicals. Unlike the decisions and chemical applications on agricultural fields, however, those associated with residential lawns are more often related to comfort and aesthetic than economic rationales (*6*). The front lawns of most homes in urban and suburban areas abut directly onto an impervious surface (i.e. streets) so that even a moderate rainfall may transport significant quantities of turf care chemicals into the nearest water body (*7*). There are also external forces influencing homeowner lawn maintenance activities, including the voices of such "experts" as horticultural writers, environmental custodians and lawn care professionals.

For all these reasons, the use of turf care chemicals on residential lawns has both scientific and public policy dimensions, and neither are adequately understood at this time. To help address this problem, a collaborative interdisciplinary project was mounted with the financial assistance of a United States Environmental Protection Agency (USEPA) Science to Achieve Results (STAR) Research Grant (EPA# 020019-01, 1999), to examine homeowner beliefs, values and socioeconomic status as they relate to the application loads and ecological impacts of turf care chemicals in aquatic ecosystems in suburbanized watersheds in Peachtree City, GA.

Residential Lawns

For homeowners, residential lawns represent not only a desire, but also the ability to grow and tend turf grasses. According to the 1991-1992 National Gardening Survey, 62% of all U.S. households engaged in the care of lawns averaging 1/3 acre and covering in total an estimated 20 million acres. The estimated annual retail sales of residential lawn care products and equipment was $6.9 billion, and turf and lawn maintenance was estimated to be a $25 billion per year industry (*3, 8*).

Lawns as the setting for the home hold a special place in American life. Separateness is the essential characteristic of suburban life, as it expresses the Jeffersonian ideal of agrarian independence, the source of civic virtue (*8, 9*). Suburban tract housing also rests on the civic and political belief that every family can, and should, own its own house with a yard, so that the single-family house has come to represent a fundamental American democratic ideal (*8-10*).

The rugged individualism of this ideal is contrary to the needs and realities of urban living, and this introduces the need to consider the role of experts. As urban development accelerated in the U.S., many jurisdictions adopted the Basic Property Maintenance Code to regulate the property maintenance activities of homeowners to protect public health, safety and welfare (*3*). The ordinances and regulations are enforced by custodians of public welfare, including local health and environmental departments.

The professional lawn care industry, in turn, promotes a view that residential lawns provide aesthetic, environmental and economic benefits. For example, the Professional Lawn Care Association of America (PLCAA) in Marietta, GA makes the case (e.g., *The ABCs of Lawn & Turf Benefits*) that a healthy turf provides oxygen for people to breathe, has the cooling effect of air conditioning, traps dust and dirt, and adds property value to real estate (*11*). Horticulture writers build from a house-in-the-park aesthetic to portray the lawn as an extension of the home that symbolizes both good domestic management (i.e., a neatly kept lawn indicates a family's life is in order) and the conquest of nature reduced to "...delightful 'accidents'" in the intensively managed lawn (*12*).

Peachtree City

Peachtree City is a suburban community in southwestern Fayette County, Georgia, within the metro-Atlanta area and formally chartered March 9, 1959 (*13*). The city is the product of a joint real estate venture established in 1957 that involved several local developers led by Joel Cowan. The corporation bought 15,000 rural acres centered on the main east-west and north-south state transportation arteries, respectively Georgia State Highway 54 and Georgia State Highway 74 (*14*).

At the time development of Peachtree City began, Atlanta was the 22nd largest metropolis in the U.S., with thriving industrial, merchandising, wholesaling, medical service and office operations centered on well-developed transportation lines that connected the city with the rest of the country. Atlanta has continued to grow since the late 1950s, and currently heads the "Top 20" list of land-consuming metropolitan areas in the nation (eleven of the Top 20, of 312 total, are in the southeast) (*15*). Between 1982 and 1997, Atlanta experienced an 81% increase in area and a 40% increase in population.

Table I summarizes the development history of Peachtree City. The rapid growth in the 1970s and 1980s was a response to the $20 million spent on expanding Hartsfield International airport. The airport is within easy access of Peachtree City residents and a prime source of employment; by 1990, 60% of Peachtree City families had at least one household member working in the airline industry. Hartsfield is now the busiest airport in the world, and as of 2005, was half-way through a 10-year, $1+ billion dollar expansion.

Peachtree City was established on "New Town" development principles in response to the unplanned growth of Atlanta (*16*). As elsewhere, the "New Town" movement was a response to post-World War II suburban sprawl, and the diminished quality of life this meant for urban dwellers. The movement focused on regional planning, decentralization away from existing overcrowded urban cores and limiting the size of local development and redevelopment activities through strict housing standards. The standards typically included strong aesthetic control and preservation of the natural environment (*17*). "New Towns" sought to ensure the benefits of living in the country while retaining the amenities of an urban existence (*18-22*).

Peachtree City is organized into four super block "villages." Each super block is built around a lifestyle choice with recreation as the dominant theme – two are built around golf courses, one is built around equestrian trails, and one is built around wetlands and "passive recreation." Villages consist of multiple neighborhood clusters of 1,500-2,000 families, with high- and low-rent properties originally planned throughout the city's neighborhoods to minimize the formation of socioeconomic or ethnic enclaves (*16*). As of 2000, Peachtree City contained 164 neighborhoods, 8,201 properties and a total population of 31,580.

In keeping with "New Town" principles, Peachtree City neighborhoods were built following strict ordinance codes so that they would have different characters without being internally homogeneous. Streets within neighborhoods are curvilinear and designed to accommodate local traffic only, while most side streets terminate in cul-de-sacs. Collector streets direct traffic around neighborhood clusters toward "maxi-centers" where residents have easy access to shopping, recreational facilities, and K-12 schools.

Table I. Peachtree City Development History

Date	*Owner*	*Neighborhoods Built*	*Properties Developed*	*Population*
1960	Bessemer Securities of New York	13	614	794
1970	Bessemer Securities of New York	37	1,631	6,500
1980	Equitable Life Insurance Company	108	5,154	21,000
1990	Pathway Communities	162	8,189	31,500
2000	Pathway Communities	164	8,201	31,580

SOURCES: Ewing 1991 (*12*); Reinberger 2002 (*22*); PTC Engineering Office 2002; United States Census 2000

The city presently contains three golf courses, two lakes, a 2,200-seat amphitheatre, over 70 miles of golf cart paths, a tennis center, an indoor swimming complex, a BMX track, a skate park, and a multitude of other neighborhood parks and recreation facilities. This relatively large recreational surface for an urban area was possible because Peachtree City developers set aside 3,000 acres as open space at the beginning of the development (*13, 14*). The extensive common space also provides for chance social interaction, and creates a feeling of community for many residents.

Peachtree City has achieved many of the "New Town" ideals of creating a sense of community, providing many cultural and recreation opportunities within the city, retaining a desirable aesthetic throughout the city and preserving open space that is easily accessible to residents. In 1976, *Better Homes and Gardens* and the National Association of Home Builders presented Peachtree City with their Grand Award for the genius of its planning and the quality of its development (*13*). The city has nevertheless fallen short of other "New Town" ideals. Much of the multi-family and lower-income housing planned was never developed. Peachtree City is also not self-sufficient, since many residents work and shop outside the city limits. Finally, the tight clusters of homes that are a trademark of "New Towns" were quickly abandoned in favor of larger lots.

Methods

The premise of the socioeconomic component of this research is that beliefs, values and practices about lawn and lawn care are structured in accordance with cultural models. Beliefs are what people think the world is like, values are the guiding principles of what is moral, desirable or just, and practices are manifest behaviors (*24, 25*). Cultural models are presupposed, taken-for-granted representations of the world that are widely shared by the members of a society or social group. They constitute what an individual needs to know in order to behave as a functioning member of their society or social group. Like formal theories, cultural models rest on abstractions that are applicable to different, yet analogous, situations to enable predictions and guide behavior. They are seldom formulated as explicit declarative knowledge, but rather as implicit knowledge embedded in words (*25-27*). Our prediction is that homeowners hold group-characteristic cultural models about lawns and lawn care.

The socioeconomic research followed a quasi-experimental design to explore and explain the relationship between beliefs, values and practices of lawn care. We used a nested, mixed methods approach, and deliberately sampled for heterogeneity (*24, 28-30*). The treatment was natural differences in lawn and lawn care beliefs, values and practices while the experimental units were homeowners (*24, 29, 30*). (We also interviewed 25 "experts" of various disciplines about lawn care in Peachtree City from a public welfare or tradesman perspective, but do not discuss these results in this article.)

Homeowners were selected using spatially-explicit criteria from neighborhoods within three defined socioeconomic groups, based on mean property value: High Income \$438,336; Medium Income \$259,573; and Low Income \$72,319. A door-to-door survey was conducted with 47 homeowners, with simultaneous observations on the structural elements of the lawn rated along dimensions of care.

Twenty four homeowners were then selected for in-depth interviews from neighborhoods co-located with aquatic sampling site watersheds. Homeowners were interviewed in person using a protocol of questions that addressed beliefs, values and lawn care practices. A subset of seven homeowners also maintained lawn diaries in which they recorded their maintenance activities and expenses over a 2-year period.

Subsequently, eleven additional homeowners were selected from four neighborhoods for explanatory interviews, to confirm the cultural model of lawn care that emerged from the set of 24 interviews. These participants were also asked to rank a series of 13 images of residential properties from least well-maintained to most well-maintained, and to explain their choices.

In-depth homeowner interviews (35 total: 24 exploratory and 11 explanatory) were taped, transcribed verbatim and then processed in QSR NVivo 2.0 using etic codes following the principles of grounded theory (5, 31). Data from the door-to-door survey and the ranking exercise were analyzed using matrix algebra, with similarities and differences determined using exploratory and categorical procedures (*24, 32-34*).

Results from in-depth interviews, the door-to-door survey, and the ranking exercise were used to develop a fixed-form economic survey that focused on homeowner decisions affecting lawn care and environmental quality in the greater Atlanta area. The survey was pre-tested, then conducted by telephone by the University of Georgia Survey Research Center in November and December of 2003. A total of 500 useful responses (200 in Peachtree City and 300 in the metro Atlanta area) were obtained from homeowners residing in neighborhoods matching those where environmental and socioeconomic research had been conducted. Particular attention was given to homeowner decisions about chemical use, fertilizer use, irrigation choices, the use of professional lawn care services as well as the time and money spent on lawn care and maintenance.

Results

The door-to-door survey raised interesting questions about what mattered to whom and where they lived within the city. High income respondents labeled toxicity and price as the forces driving their choices of lawn care products, and viewed environmental friendliness and convenience of use as insignificant. Low income respondents identified price as well as convenience as important, while toxicity was not important. Separation between socioeconomic groups was less pronounced when questioned about the relative importance of lawn issues. "Having a lawn that looks as good as the neighbor's lawn" was very important to all three socioeconomic groups. High and Low socioeconomic groups also had in common the importance of a lawn that was green all year, but this issue was not important to the Middle socioeconomic group.

The conclusion derived from these and the other results of this survey is that homeowner choices are not monolithic. Socioeconomic groups had different concerns when faced with real world choices that impacted their immediate economic wellbeing and their surrounding natural environment.

The in-depth interviews revealed that homeowners identified pride of ownership, respect for neighbors and pride of place as the primary motivations for maintaining their lawns. They also emphasized the importance of context for evaluating the quality of the lawn maintenance efforts of other neighborhood residents. A lawn was evaluated for consistency with its surroundings, and in the overall flow and balance of the landscape throughout a neighborhood. Homeowners also described how neighborhood social process – peer pressure of various kinds – affected their decisions about landscape maintenance. Within-neighborhood participants were highly consistent in their overall ranking of landscapes, but differences between neighborhoods were significant (Figure 1).

Peachtree City residents generally preferred working to improve and maintain their plants and gardens to working on their lawns. Seventy-three percent reported that caring for their gardens was more of a pleasant hobby than an unpleasant chore, while only 45% reported that caring for their lawns was more of a pleasant hobby than an unpleasant chore. Fifty-six percent of respondents reported paying someone to work on their lawns within the past year. Thirty percent used a chemical applicator (e.g., TruGreen® ChemLawn®), paying an annual average of $647; 57% hired a landscape maintenance company for an annual average of $1064; 11% hired a gardener or handyman for an annual average of $795; while 17% hired a neighborhood teenager for an annual average of $414. The need to maintain neighborhood standards was a more important determinant of lawn care labor choices than of choices about working with plants and gardens.

The lawn maintenance activities most likely associated with environmental damage were the purchase and application of herbicides, pesticides, and synthetic fertilizers. The average Peachtree City household spent 20 hours and

$103 per year applying herbicides, pesticides and synthetic fertilizers. Of the 48% of households applying non-zero amounts of these products, the average was 36 hours and $215. Biological pest controls of low-toxicity insecticides were used by 23% of Peachtree City households for an average of six hours at a cost of $44 versus two hours and $51 for the use of biological pest controls.

Among providers of professional lawn care services, the chemical applicator companies were by far the most likely to perform tasks involving the application of pesticides, herbicides, and fertilizers. Eighty-one percent of homeowners hiring these companies had them apply herbicides and fertilizers, and 57% reported that they applied insecticides. Among landscape maintenance companies, 60% applied fertilizers, 42% applied herbicides and 32% applied insecticides. Among other paid service providers (gardeners and neighborhood teenagers), only 7% applied fertilizers and insecticides and none applied herbicides.

Figure 1. Ranked residential lawn preferences: A) New, Low Income (NL) most preferred; B) NL least preferred; C) Old, High Income (OH) most preferred; D) OH least preferred. The NL neighborhood is on a cul-de-sac and has a single straight street with small lots; there is little variation in topography and trees tend to be small since this is a young neighborhood. The OH neighborhood is on a meandering loop street with large lots; the topography is extreme and trees are mature making lawn maintenance difficult for most residents.

Income and education provided no explanatory power for individual components of the lawn care of Peachtree City residents, including hours or money spent on the application of lawn chemicals. Income and education also had no explanatory power for whether Peachtree City residents choose to hire chemical applicators or landscape maintenance companies for lawn care work. However, income did have a positive relationship with money spent on pesticides, herbicides and fertilizers, and a highly statistically significant relationship with total expenditures on lawn and garden maintenance. The average total expenditure for Peachtree City residents was $921 per year, and each additional hundred dollars of income was associated with about $1 of additional expenditure on lawns and gardens.

Water usage was a critical factor of the environmental impact of lawn care, but not as a consequence of lawn care chemical impact. In Georgia, increasing surface water flows are important to ecosystem health during the summer months. Fifty-one percent of Peachtree City households spent time watering their lawns, averaging 61 hours per year. Thirty-five per cent of Peachtree City households had automatic underground sprinkler systems; 9% had automatic drip systems; and 36% used a hose or watering can on their lawns.

Peachtree City households were aware of the importance of watering their lawn during the cooler part of the day, and only 4% reported watering between 10 AM and 4 PM. Fifty four per cent watered in the morning before 10 AM and 42% watered after 4 PM. Residents reported that their watering decisions were most strongly influenced by knowledge of drought and watering restrictions, closely followed by concern about the appearance of their lawns. Half of the respondents watered less and changed their watering to cooler parts of the day because of their awareness of Georgia's drought conditions. Twenty-two percent reported that they increased the amount they watered because drought conditions increased their lawn's need for water. Twenty-one per cent reported making changes in the composition of their yard to reduce their need to water.

Discussion

In this USEPA-sponsored collaborative research, we assessed how homeowner beliefs, values and socioeconomic status determined application loads and the ecological impacts of turf care chemicals on suburban aquatic ecosystems. Because lawns and the various practices associated with their upkeep have typically been studied through aggregate statistical methods or from the normative perspective of landscape design, virtually nothing is known about how social, economic or aesthetic forces structure the behavior of homeowners. For example, some "simple" yet important questions have never been asked: How do residents of socioeconomically distinct neighborhoods choose between lawn additives?

The beliefs, values and practices about lawn and lawn care, whatever else they may represent, are structured in accordance with cultural models that motivate practical behavior in line with the authority and expertise they are invested with, or the intrinsic persuasiveness with which they are presented. Perhaps the only recourse for curbing the rugged individualism of suburban lawn care practices is through carefully enforced regulations such as those contained in the basic Property Maintenance Code published by the Building Officials and Code Administrators International (*3*).

For homeowners, residential lawns represent not only a desire, but also the ability to grow and tend turf grasses. The desire developed from exposure to the landscape aesthetic of upper class suburbs and golf courses developed during the middle part of the 20th century. The ability in the hotter regions of the country, including the Southeast, only became possible after the development of hardy turf varieties in the late 1950s and early 1960s (*3, 8*). Lawns, in short, represent time, money and labor for the homeowner.

Yards and lawns are important to Peachtree City homeowners, and they expend labor and money in varied ways on the setting for their homes. While they use products to care and maintain their yards and lawns that have the potential for environmental damage, the activities are not universally practiced. Professional applicators of insecticides, herbicides and fertilizers and lawn maintenance companies are prevalent, but household labor remains significant.

Peachtree City homeowners signal pride and respect to one another and to a greater audience through their residential landscape maintenance practices; they strive for consistency at the residential level to ensure the balance and flow of a neighborhood-level landscape. Homeowners clearly influence each other in profound ways not only through their maintenance practices, but also through neighborhood social processes.

While our research in Peachtree City examined homeowners, it is really about group process and what is termed "cognition in the wild" (*35-37*). Managing lawns and applying lawn care chemicals represent individuated and unregulated activities for which we go beyond idiosyncratic and group-bound worldviews in order to bring the results of our inquiry in line with the needs of decision- and policy-makers. If it can be determined how the meaning systems associated with suburban lawns are acquired, organized and used, then we are in a position to create incentives to elicit alternative choices that leverage social action in ways that do not perpetuate extant problems.

Policies are designed to influence the choices made by vast numbers of people who collectively produce the desired outcome. Arguing that lawns are beyond criticism or reform because they are naturalized social practices (*38*), or that they represent the failure to envision alternative technical solutions (*3*) fails to address fundamental issues pertaining to how urban and suburban landscapes can remain habitable. We live in a world of use, and we need to consider how use can be organized and structured for the long-term mutual benefit of nature and humans.

Urbanization is a dominant and worldwide process, with important effects on regional landscapes. In the United States, the conversion of wilderness and agricultural-use lands to urban/suburban-use lands is faster than the rate of population growth. This creates challenging problems for policy makers and planners, and yet the attention given to urban/suburban ecosystems is insignificant by comparison to the effort devoted to "natural" ecosystems (*4, 39-41*). The staggeringly complex problems associated with urbanization – trash disposal, automobile use, home maintenance, etc. – are typically approached monolithically, as if the systems themselves had volition. It is nevertheless the choices made by the many individual residents of urban areas that collectively drive the large scale processes that characterize such systems. The very ordinariness of the multiple daily decisions makes the processes easy to overlook, and so we miss the significance of how the mundane combines into large effects (*4, 41*).

Most resource management efforts target the average within large scale geographical areas, e.g., Total Maximum Daily Loads (TMDLs). In these efforts, however, the lack of attention to a nuanced, scalar analysis of interactions between the socioeconomic and biophysical domains causes system managers to ignore potential disproportionate contributions that may be driving the output of the system. Who adopts or why they adopt may not be as important as where and when certain behaviors occur or need to occur (*42*). Our research in Peachtree City and the greater Atlanta area clearly shows the importance of accounting for the scale-specific biophysical setting of human behavior. It also indicates that any natural resource management program that strives for effectiveness and efficiency needs to be based on an interdisciplinary partnership of social and biophysical scientists.

Acknowledgements

This manuscript is based on research supported by a US Environmental Protection Agency Science to Achieve Results (STAR) Program grant to Kevin Armbrust (Grant# 020019-01 1999). Much of the effort and many of the insights were provided by the students who worked on different phases of the research: Cheryl McClary, Ken Goodman, John Primo and Nathan Piekielek. Any opinions, findings, conclusions, or recommendations expressed in the material are those of the authors and do not necessarily reflect the views of the US Environmental Protection Agency.

References

1. Henderson, S. P. B.; Perkins, N. H.; Nelischer, M., Residential Lawn Alternatives: A Study of Their Distribution, Form and Structure. *Landscape and Urban Planning,* **1998**, 42, (2-4), 135-145.
2. Blanco-Montero, C. A.; Bennett, T. B.; Neville, P.; et al., Potential Environmental and Economic Impacts of Turfgrass in Albuquerque, New Mexico (USA). *Landscape Ecology,* **1995**, 10, (2), 121-128.
3. Bormann, F. H.; Balmori, D.; Geballe, G. T.; Vernegaard, L., *Redesigning the American Lawn: A Search for Environmental Harmony*. 2nd ed.; Yale University Press: New Haven, 2001.
4. Robbins, P.; Polderman, A.; Birkenholtz, T., Lawns and Toxins: An Ecology of the City. *Cities,* **2001**, 18, (6), 369-380.
5. Feagan, R. B.; Ripmeester, M., Contesting natural(ized) lawns: A Geography of Private Green Space in the Niagara Region. *Urban Geography* **1999**, 20, (7), 617-634.

6. Templeton, S. R.; Zilberman, D.; Yoo, S. J., An Economic Perspective on Outdoor Residential Pesticide Use. *Environmental Science and Technology* **1998**, 32, (18), 416A-422A.
7. Smith, A. E.; Bridges, D. C., Potential Movement of Certain Pesticides Following Application to Golf Courses. In *Herbicide Metabolites in Surface Water and Groundwater*, Meyer, M. T.; Thurman, E. M., Eds. ACS Press: 1996.
8. Jenkins, V. S., *The Lawn: A History of an American Obsession*. Smithsonian Institution Press: Washington, DC, 1994.
9. Kemmis, D., *Community and the Politics of Place*. University of Oklahoma Press: Norman, 1990.
10. Borgmann, A., *Crossing the Postmodern Divide*. The University of Chicago Press: Chicago, 1992.
11. *ABCs of Lawn and Turf Benefits*, Professional Landscape Network. www.LandcareNetwork.org.
12. Nesmith, L., A New State of Mind. *Southern Living,* **1999**, 34, (3), 140-143.
13. Ewing, R., *Profiles of Five Successful New Communities*. Glatting Lopez Kercher Anglin, Inc.: Washington D.C., 1991.
14. Riley, 1965.
15. Rivers, A., 2002, 7.
16. Phipps, L. C., PTC Master Development Plan, 1972.
17. Morley, J. C., *The New Town*. Building Research Institute, National Research: Washington, DC, 1966.
18. Conklin, W. J., *Planning Approaches to New Towns*. Building Research Institute, National Research: Washington D.C., 1966.
19. Finley, W. E., *New Towns of the Future*. Building Research Institute, National Research: Washington D.C., 1966.
20. Gladstone, R.; Wise, H. F., New Towns Solve Problems of Urban Growth. *Public Management,* **1966**, 48, 128-139.
21. Lemkau, P. V., *Human Factors in the New Town*. Building Research Institute, National Research: Washington D.C., 1966.
22. Simon, R. E., *Problems of the New Town Developer*. Building Research Institute, National Research: Washington D.C., 1966.
23. Reinberger, M. In *Evolution of Landscape Design in Peachtree City, Georgia*, Society for Applied Anthropology Annual Meeting, Atlanta, GA, 2002; Atlanta, GA, 2002.
24. Kempton, W.; Boster, J. S.; Hartley, J. A., *Environmental Values in American Culture*. The MIT Press: Cambridge, MA, 1995.
25. Quinn, N.; Holland, D., Culture and Cognition. In *Cultural Models in Language & Thought*, Holland, D.; Quinn, N., Eds. Cambridge University Press: Cambridge, 1987; pp 3-40.
26. D'Andrade, R., *The Development of Cognitive Anthropology*. Cambridge University Press: Cambridge, UK, 1995.
27. Kempton, W., Two Theories of Home Heat Control. In *Cultural Models in Language & Thought*, Holland, D.; Quinn, N., Eds. Cambridge University Press: Cambridge, 1987; pp 222-242.

28. Creswell, J., *Research Design: Qualitative, Quantitative, and Mixed Methods Approaches.* 2nd ed.; Sage Publications: Thousand Oaks, CA, 2003.
29. Cook, T. D.; Campbell, D. T., *Quasi-Experimentation: Design and Analysis for Field Settings.* Rand McNally: Chicago, 1979.
30. Johnson, J. C., Research Design and Research Strategies. In *Handbook of Methods in Cultural Anthropology,* Bernard, H. R., Ed. Altamira Press: Walnut Creek, CA, 1998; pp 131-171.
31. Strauss, A.; Corbin, J., *Basics of Qualitative Research: Grounded Theory, Procedures and Techniques.* Sage Publications: Newbury Park, CA, 1990.
32. Borgatti, S. P., *Anthropac 4.0 Reference Manual.* Analytic Technologies: Columbia, 1993.
33. Noreen, E. W., *Computer-Intensive Methods for Testing Hypotheses.* John Wiley: New York, 1989.
34. Simon, J. L., *Resampling: The New Statistics.* Resampling Stats, Inc.: Arlington, VA, 1997.
35. Holland, D. C.; Eisenhart, M. A., *Educated in Romance: Women, Achievement and College Culture.* Chicago: University of Chicago Press, 1990.
36. Hutchins, E., *Cognition in the Wild.* MIT Press: Cambridge, MA, 1995.
37. Kempton, W., Cognitive Anthropology and the Environment. In *New Directions in Anthropology and Environment*, Crumley, C., Ed. Altamira: Walnut Creek, CA, 2001; pp 49-71.
38. Feagan, R. B.; PRpimeester, M., Contesting Naturalized Lawns: A Geography of Private Green Space in the Niagara Region. *Urban Geograph,y* **1999**, 20, (7), 617-634.
39. Botkin, D. B.; Beveridge, C. E., Cities as Environments. *Urban Ecosystems* **1997**, 1, 3-19.
40. Grimm, N. B.; Grove, J. M.; Pickett, S. T. A.; Redman, C. L., Integrated Approaches to Long-Term Studies of Urban Ecological Systems. *BioScience* **2000**, 50, (7), 571-584.
41. Pickett, S. T. A.; Cadenasso, M. L.; Grove, J. M.; Nilon, C. H.; Pouyat, R. V.; Zipperer, W. C.; Costanza, R., Urban Ecological Wystems: Linking Terrestrial, Ecological, Physical and Socioeconomic Components of Metropolitan Areas. *Annual Review of Ecology and Systematics,* **2001**, 32, (2001), 127-57.
42. Nowak, P.; Bowen, S. K.; Cabot, P. E., Disproportionality as a Framework for Linking Social and Biophysical Systems. *Society and Natural Resources,* 2006 (in press).

Chapter 3

Effects of Turf Pesticides on Aquatic Invertebrates

Jay P. Overmyer

Department of Entomology, University of Georgia, Athens, GA 30602

A review of the literature on the effects of turf pesticides on aquatic invertebrates was conducted. Of the 51 pesticides registered for use on turf, information was summarized for seven insecticides, nine herbicides and three fungicides. In general, turf insecticides pose the greatest threat to aquatic invertebrates, indirect effects of herbicides are more detrimental than direct effects, and fungicides have potential endocrine disrupting effects.

Aquatic ecosystems have the potential to be highly impacted by chemicals used to protect turf from pests. Turf pesticides enter waterways primarily through runoff from rain events and are commonly detected in streams and rivers (*1*). Several chemicals, primarily insecticides, have been detected at concentrations exceeding aquatic life exposure criteria (*1*). Thus, organisms inhabiting streams receiving turf runoff might be adversely affected.

Invertebrates are key components of aquatic ecosystems. They aid in critical ecosystem functions such as organic matter processing and transport, and allow the transfer of energy from basal resources to higher trophic level organisms. In addition, invertebrates are often highly sensitive to pesticides, and can serve as bioindicators of contamination and potential loss of ecosystem integrity.

In this chapter, the literature on the effects of turf pesticides on aquatic invertebrates is reviewed. Literature was gathered for pesticides currently registered for use on turf in the state of Georgia (Table I) (*2*). Many of the chemicals listed are also registered for use on other commodities. Studies were cited based on the effects of the active ingredients, not the crop it is used for. Thus, many of the studies cited are not specifically related to the chemical's use on turf. The works compiled in this review are by no means all encompassing, since pesticides in which there was only toxicity test data available were not included. Also, studies using similar indicator organisms or which produced redundant information were selectively included. However, the literature summarized in this chapter gives us a good "flavor" of the type of work that has been conducted and potential areas where further research is needed.

Insecticides

Turf insecticides have received considerable attention in relation to their effects on non-target invertebrates because insecticides are primarily neurotoxicants that bind to receptors in both target and non-target organisms. Many organophosphate insecticides registered for turf have been discontinued and replaced with chemicals with higher specificity for insect receptors. Thus, the newer insecticides used on turf are not highly toxic to all invertebrate species. However, most aquatic insect species are extremely sensitive to insecticides, and more research is needed in assessing their effects on these non-target species.

Carbaryl

Carbaryl is a carbamate insecticide that has been used for control of insect pests on turf for many years. Thus, extensive research has been conducted with this chemical on target and non-target invertebrates. Federle and Collins (*3*) assessed the toxicity of carbaryl to the backswimmer, *Notonecta undulata*, and the water beetle, *Peltodytes* sp. The 96-hr median lethal concentration (LC_{50}) values for *N. undulata* and *Peltocytes* sp. were 0.2 and 3.3 mg/L, respectively. Carbaryl was shown to be moderately toxic to the freshwater bivalve, *Corbicula striatella,* with a 96-hr LC_{50} of 5.1 mg/L (*4*). Acute and chronic exposures to this species produced a depletion in ascorbic acid in the mantle, foot, gill, digestive gland and whole body (*5*). The toxicity of carbaryl formulated as Clean Prop® was also tested with field-collected macroinvertebrates from Pacific Northwest streams (*6*). The 96-hr LC_{50} values ranged from 11.1 to 61.0 µg/L for the mayfly, *Cinygma* sp., and the caddisfly, *Psychoglypha* sp., respectively. The estimated hazard concentration to 5% of the theoretical benthic community was determined to be between 0.43 and 0.66 µg/L. Acute toxicity and genotoxic effects of carbaryl (Sevin®) were assessed with glochidia of the freshwater mussel, *Utterbackia imbecillis* (*7*). The 24-hr LC_{50} was reported to be 7.9 mg/L, but no genotoxic effects were observed at ¼ and ½ the no observed effects concentration (NOEC) of 3.49 mg/L. Overmyer et al. (*8*) assessed the toxicity of carbaryl individually and in mixtures with organophosphate insecticides using black fly larvae, *Simulium vittatum* IS-7. The 48-hr LC_{50} was determined to be 23.72 µg/L. In mixtures, the binary combination of carbaryl + malathion and the tertiary combination of carbaryl + chlorpyrifos + malathion, produced greater than additive toxicity while the binary combination of carbaryl + chlorpyrifos was additive.

Table I. Pesticides Registered for Use on Turf in the State of Georgia, USA (2004)

Insecticides	*Herbicides*	*Fungicides*
Acephate	Atrazine	Azoxystrobin
Bifenthrin	Benefin	Cloroneb
Carbaryl	Bensulide	Clorothalonil
Cyfluthrin	Bentazon	Etridiazol
Fipronil	Clethodim	Fenarimol
Imidacloprid	2,4-D	Flutolanil
Lambda-cyhalothrin	Dicamba	Fosetyl-Al
	Diclofop	Iprodione
	Dithiopyr	Mancozeb
	Ethofumesate	Mefenoxam
	Fenoxaprop	Myclobutanil
	Imazaquin	PCNB
	Isoxaben	Polyoxin-D
	MCPP	Propamocarb hydrochloride
	Metribuzin	Propiconazole
	Orzalin	Thiophanate methyl
	Oxadiazon	Triadimefon
	Pendimethalin	Trifloxystrobin
	Prodiamine	Vinclozolin
	Pronamide	
	Sethoxydim	
	Simazine	
	Triclopyr	

SOURCE: 2004 Georgia Pest Management Handbook, Commercial Edition

The toxicity of carbaryl to the sea urchin, *Pseudechinus magellanicus,* was determined to be life-stage specific, with the blastula and gastrula stages being more sensitive than the prism and pluteus (*9*). Lower toxicity in the late life-stages was thought to be related to an increase in detoxification processes by the cytochrome oxidase system. Early life-stages of the damsel fly, *Xanthocnemis zealandica*, were also more sensitive to carbaryl (*10*). The 48-hr LC_{50} values for second and thirteenth instar nymphs were 156.6 and 760 μg/L, respectively. The egg stage was determined to be the least sensitive stage, with reduced hatching success apparent at 600 μg/L.

Water quality parameters have been shown to influence the toxicity of carbaryl. Temperature and pH were shown to affect the toxicity of carbaryl to the midge, *Chironomus riparius* (*11*). The highest toxicity was observed at pH 4 and 30°C (EC_{50} = 61 μg/L) and the lowest toxicity was observed at pH 4 and 6 at 10°C (EC_{50} = 133 μg/L). The amount of organic matter present in the water can also influence the toxicity of carbaryl. Overmyer et al. (*12*) showed that concentrations of food in the water ≥ 150 mg/L significantly increased the toxicity of carbaryl to *S. vittatum* IS-7.

Several studies have assessed sublethal effects of carbaryl exposure. The feeding rate of the marine mussel, *Mytilus edulis*, was reduced in a dose dependent manner while being exposed to carbaryl at concentrations ranging from 1 to 9 mg/L (*13*). The reduced feeding rate was primarily related to the narcotic effect of carbaryl on the mussels, with a minor contribution due to neurotoxicity as a result of reduced acetylcholinesterase activity. Predator-prey interactions were studied using southern leopard frog tadpoles, *Rana sphenocephala*, as the prey and the red-spotted newt, *Notophthalmus viridescens*, as the predator (*14*). Predation rates were affected when either the predator or prey was exposed to carbaryl at 2.5 mg/L, but not when both were exposed simultaneously. Glycogen, pyruvate and lactate levels and lactate dehydrogenase activity were significantly altered in hepatopancreas and ovitestis tissues of the freshwater snail, *Lymnaea acuminate*, after a 96-hr exposure to carbaryl at concentrations ≥ 3 mg/L and ≥ 6 mg/L, respectively (*15*). Seuge and Bluzat (*16*) showed a decrease in fecundity in the freshwater snail, *L. stagnalis*, exposed to carbaryl at concentrations as low as 1 mg/L. Water fleas, *Moina micrura*, exposed to a carbaryl concentration of 30 μg/L (¼ the 24-hr LC_{50}) showed reduced growth, egg production, growth coefficient, and intrinsic rate of increase (*17*). Damselflies, *Xanthocnemis zealandica*, exposed to 100 μg/L carbaryl had reduced emergence success and greater fluctuating asymmetry in wing length (*18*).

The effects of carbaryl on zooplankton and phytoplankton community structure were assessed by Hanazato and Yasuno (*19, 20*) and Havens (*21*). Mesocosms treated with 1 mg/L carbaryl showed rapid decline in zooplankton and *Chaoborus* sp. larvae (*19*). The cladoceran population recovered rapidly and dominated the zooplankton community in the absence of the predator *Chaoborus* sp. No direct effects of carbaryl were observed in the phytoplankton community; however, the phytoplankton community structure changed following the change in the zooplankton community. An increase in phytoplankton biovolume and a decrease in zooplankton biomass were observed in mesocosms exposed to a range of carbaryl concentrations (*21*). Cladocerans were shown to be the most sensitive to carbaryl, and copepods were most tolerant. Hanazato and Yasuno (*20*) also showed that the timing of carbaryl exposure can influence the recovery of zooplankton communities. Populations of the rotifer, *Keratella valga*, exposed to carbaryl during the increasing phase continued to increase after exposure, while those exposed during the declining phase did not recover. The authors also suggest that other factors, such as temperature, competitive interactions among zooplankters and trends in zooplankton populations are important in determining community structure. Artificially colonized mesocosms dosed with 0.51 mg/L carbaryl (Sevin®) exhibited reduced species richness, reduced predator biomass and increased large herbivore biomass (*22*). However, species responses within functional groups were variable.

A discriminant analysis was conducted by Passy et al. (*23*) to determine environmental variables that predict diatom, macroinvertebrate and fish community structure. Carbaryl was shown to be a significant variable in determining macroinvertebrate groupings. Effects of aerial applications of carbaryl formulated as Sevin-4-Oil® on invertebrate drift in streams were studied by several authors (*24-26*). All reported increases in invertebrate drift after applications; however, Beyers et al. (*24*) suggested that biological significance of the increased drift was minimal. Carbaryl was also identified as the major toxicant in water samples from one site in the San Joaquin River Delta through toxicity identification evaluations using *Ceriodaphnia dubia* as the test organism (*27*).

Acephate

Acephate is the only remaining organophosphate insecticide still registered for use on turf. It has good insecticidal properties but is a weak inhibitor of acetylcholinesterase, making it much safer than many other organophosphates. Acephate (Orthene®) was shown to be moderately toxic to adult backswimmers and water boatmen , two pond-dwelling insects with reported 24-hr LC_{50} values of 10.4 mg/L and 8.2 mg/l, respectively (*28*). The author also reported that acephate is a weak cholinesterase inhibitor in insects and that the metabolite of acephate, methamidophos, is a much stronger inhibitor. Thus, acephate needs to be bioactivated to elicit toxic effects (28). Significant cholinesterase inhibition was measured in the freshwater mussel, *Elliptio complanata,* at $\geq$ 1.3 mg/L acephate; however concentrations as high as 320 mg/L produced no significant cholinesterase inhibition in the Asiatic clam, *Corbicula fluminea* (*29*). Whole body acetylcholinesterase levels were significantly reduced at concentrations $\geq$ 0.2 mg/L in the freshwater shrimp, *Paratya australiensis* (*30*). The author determined the maximum acceptable toxicant concentration (MATC) for cholinesterase inhibition in *P. australiensis* to be 0 -190 μg/L.

In a field assessment of acephate applied to streams, Rabeni and Stanley (*31*) observed no effects in invertebrate standing crop, but an increase in invertebrate drift was observed. The maximum concentration of acephate detected from water samples (140 μg/L) occurred one hour after application, with residues remaining for at least two days post treatment. Results of this study differ from those of a study by Bocsor and O'Connor (*32*), where no increase in post-spray invertebrate drift occurred in a small stream sprayed with acephate.

Fipronil

Fipronil has shown high variability in its toxicity to aquatic invertebrates. Some species were extremely sensitive while others were quite tolerant. Sensitivity to this insecticide is related to its affinity for the γ-aminobutyric acid (GABA) receptor in the organism. Fipronil appears to have a strong affinity for GABA receptors in aquatic insects. This is not surprising, as insects are typically the target species for this chemical. Overmyer et al. (*33*) showed that black fly larvae, *Simulium vittatum* IS-7, were highly sensitive to fipronil. Median lethal concentrations (LC_{50}) during 48-h exposures ranged between 0.19

and 0.29 μg/L. Chironomid and mosquito larvae were also shown to be sensitive to fipronil, with 48-h LC_{50} values ranging from 0.42 μg/L for *Chironomus crassicaudatus* to 0.87 μg/L for *Culex nigripalpus* (*34*). Another midge species, *C. annularis*, had a 48-h LC_{50} of 5.6 nM (*35*). Other (non-insect) species highly sensitive to fipronil are adult and larval grass shrimp, *Palaemonetes pugio*, with reported 96-h LC_{50} values of 0.32 and 0.68 μg/L, respectively (*36*) and the mysid, *Americamysis bahia*, with a reported 24-h LC_{50} of 0.14 μg/L (*37*). Additional invertebrates tested with fipronil for acute toxicity have been the red swamp crayfish, *Procambarus clarkii*, and the white river crayfish, *Procambarus zonangulus,* with reported 96-h LC_{50} values of 14.3 and 19.5 μg/L, respectively (*38*); the water flea, *Ceriodaphnia dubia*, with a reported 48-h LC_{50} of 17.7 μg/L (*39*); and copepods, *Acanthocyclops robustus* and *Diaptomus castor*, with reported 48-h LC_{50} values of 194.2 and 7.9 nM, respectively (*35*).

The effects of fipronil and its degradation products, desthionyl fipronil and fipronil sulfide, on the life-cycle of the estuarine copepod, *Amphiascus tenuiremis*, were studied by Chandler et al (*40, 41*). In their studies, fipronil was more acutely toxic to males (96-h LC_{50} = 3.5 μg/L) than females (96-h LC_{50} = 13.0 μg/L). Concentrations as low as 0.22 μg/L significantly delayed development of both male and female copepods and significantly affected female egg extrusion. Based on a Leslie matrix population growth model, net production of copepods was depressed at a fipronil and desthionyl fipronil concentration of 0.25 μg/L and a fipronil sulfide concentration of 0.15 μg/L, compared to controls. However, this effect was not manifested in all treatments until the third generation. Effects at the copepod population level appear to stem from reproductive dysfunction in males as opposed to females (*42*). The effects of fipronil on population parameters such as birth rate, intrinsic rate of increase, generation time, death rate and doubling time were assessed with the water flea, *Daphnia pulex* (*43*). Effects in these parameters were only observed at a concentration equivalent to the approximate 48-h LC_{50} (0.03 mg/L) for *D. pulex*. Thus, impacts of fipronil at environmentally realistic concentrations should be minimal to this invertebrate species. Fipronil was also assessed as a potential endocrine disruptor in female grass shrimp, *P. pugio* (*44*). No effects were observed in egg production, vitellogenin, cholesterol or ecdysteroid concentrations at concentrations as high as 200 ng/L.

Imidacloprid

Imidacloprid is another insecticide that is highly specific to insect species. Of the few extant studies investigating the toxic effects of imidacloprid to aquatic invertebrates, aquatic insects have been shown to be the most sensitive. Song et al. (*45*) demonstrated that two mosquito species, *Aedes aegypti* and *Aedes taeniorhynchus,* were more susceptible to imidacloprid than the water flea, *Daphnia magna,* and the brine shrimp, *Artemia* sp. The 48-h LC_{50} values ranged from 0.013 mg/L for *A. taeniorhynchus* to 361.23 mg/L for *Artemia* sp. The acute toxicity of imidacloprid was also tested with black fly larvae,

Simulium vittatum IS-7 (*33*). The 48-h LC_{50} for this black fly species ranged between 6.75 and 9.54 µg/L.

Bifenthrin, Cyfluthrin & Lambda-cyhalothrin

Bifenthrin, cyfluthrin and lambda-cyhalothrin are second generation pyrethroid insecticides, which possess greater insecticidal activity than first generation pyrethroids. All three have been shown to be highly toxic to invertebrates inhabiting the water column as well as in the sediments where these pyrethroids tend to accumulate. Because the majority of the literature associated with the toxic effects of these chemicals includes all three, they will be discussed as a group rather than individually, to avoid redundancy.

As previously mentioned, bifenthrin, cyfluthrin and lambda-cyhalothrin are extremely toxic to aquatic invertebrates. Mokry and Hoagland (*46*) assessed the toxicity of these chemicals with the water fleas *Daphnia magna* and *Ceriodaphnia dubia.* The 48-h LC_{50} values for bifenthrin for both species were between 0.32 and 0.07 µg/L; for cyfluthrin, between 0.17 and 0.14 µg/L; and for lambda-cyhalothrin, between 1.04 and 0.30 µg/L. A toxicity assessment of the individual enantiomers of cyfluthrin to *C. dubia* showed that the 1*R-cis-αS* and the 1*R-trans-αS* isomers were the most toxic, with reported 96-h LC_{50} values of 0.104 and 0.214 µg/L, respectively (*47*). The 1*R-cis* enantiomer of bifenthrin was shown to be approximately 17-22 times more toxic than the 1*S-cis* enantiomer, indicating that the majority of toxicity in *cis*-bifenthrin is due to 1*R-cis*. (*47*).

The toxicity of these three insecticides in sediments was assessed using the amphipod, *Hyalella azteca* (*48*). The 10-d LC_{50} for bifenthrin was 0.18 µg/g OC (organic carbon), and the LOEC for reduced growth was between 0.08 and 0.21 µg/g OC; for cyfluthrin, the LC_{50} was 1.08 µg/g OC and the LOEC was between 0.46 and 0.77 µg/g OC; for lambda-cyhalothrin, the LC_{50} was 0.45 µg/g OC and the LOEC was between 0.23 and 0.14 µg/g OC (*48*). Heimbach et al. (*49*) assessed the effects of the cyfluthrin formulation, Baythroid®, on the communities of natural and artificial ponds. Declines in cladoceran populations were observed within hours of applying the insecticide to ponds at rates of 12.5 and 62.5 g/ha, and 2.5 and 12.5 g/ha, for natural and artificial ponds, respectively. However, cladoceran populations recovered within 2 weeks from the lowest application rate, and within 4 weeks from the highest application rate, as cyfluthin concentrations declined. The toxicity of sediments receiving runoff from agricultural areas was assessed by Weston et al (*50*). They determined that mortality observed in *Chironomus tentans* and *H. azteca* was related to the presence of pyrethroids, and that concentrations of bifenthrin and lambda-cyhalothrin in the sediments were high enough to account for this mortality. The effects of lambda-cyhalothrin (Karate®) on benthic macroinvertebrates were studied using stream mesocosms in a pulse-exposure scenario (*51*). Mesocosms were exposed to nominal concentrations of 0.10, 1.00 and 10.0 µg/L for 30 min during the first exposure and 0.05, 0.50, and 5.00 µg/L for 30 min during a second exposure. Total drift was significantly higher at all treatment levels after each exposure. The response of macroinvertebrate densities was more variable, with differences related to the time of sampling and exposure concentration.

A probabilistic risk assessment for pyrethroids was performed by Solomon et al. (*52*). Based on laboratory toxicity data, they determined the 10th centile value for bifenthrin, cyfluthrin and lambda-cyhalothrin to be 15, 12, and 10 ng/L, respectively. Thus, concentrations below these values should protect approximately 90% of the organisms exposed. In a risk assessment performed by Haith and Rossi (*53*), cyfluthrin concentrations in receiving waters were predicted not to cause serious threat to biota; however, during extreme rain events, concentrations approached or exceeded LC_{50} values for rainbow trout and water fleas.

Herbicides

The majority of pesticides registered for use on turf fall under the category of herbicides. Currently, there are 25 herbicides registered for controlling weeds and unwanted grasses on turf in the state of Georgia (*2*). The majority of these herbicides are not acutely toxic to aquatic invertebrates, especially at the concentrations commonly detected in the environment. However, many herbicides are toxic to aquatic plants such as algae and macrophytes, which could lead to indirect effects on the invertebrate community through loss of food sources or loss of oxygen due to reduced photosynthesis and decaying of dead plant material. Although there are numerous turf herbicides, only nine were located in the literature concerning effects on aquatic invertebrates.

Atrazine

Atrazine is probably the most widely studied herbicide in reference to its nontarget effects on aquatic invertebrates. This is likely related to its heavy use patterns and presence in aquatic systems throughout the world. Macek et al. (*54*) assessed the toxicity of atrazine in several invertebrate species. The midge, *Chironomus tentans*, was shown to be the most sensitive, with a 48-hr LC_{50} of 0.72 mg/L. The water flea, *Daphnia magna*, was the least sensitive with a 48-hr LC_{50} of 6.9 mg/L. The fiddler crab, *Uca pugnax*, was shown to be insensitive to atrazine, with survival effects occurring only at extremely high concentrations (10,000 mg/L) (*55*). The scud, *Gammarus pulex*, was shown to be more sensitive than the midge, *Chironomus riparius*, to atrazine (*56*). The 10-day LC_{50}s for the scud and the midge were 4.4 and 18.9 mg/L, respectively. Streit and Peter (*57*) assessed the long-term effects of atrazine in three invertebrate species, the snail, *Ancylus fluviatilis*, and the leeches, *Glossiphonia complanata* and *Helobdella stagnalis*. Dose-dependent decreases were observed in growth and egg production, while ingestion increased with increasing atrazine concentration. Results also showed that effects in the endpoints measured increased with exposure duration. The effect of salinity on atrazine toxicity was assessed using the copepod, *Euytemora affinis* (*58*). Based on NOEC and LOEC values determined from survival data, slight increases in salinity from 5 ppt to 15 ppt had a protective effect on atrazine toxicity. However, salinity at 25 ppt made atrazine more toxic to the copepod. The author suggests that *E. affinis* is more effective at metabolizing atrazine at 15 ppt than at the other two salinity concentrations. The presence of sediments has also been shown to affect atrazine toxicity. Phyu et al. (*59*) showed that the addition of sediment to an acute, 10-day static toxicity test significantly reduced the toxicity of atrazine to the midge, *C. tepperi*.

The effects of chemical mixtures containing atrazine have been investigated by several authors. Douglas et al. (*60*) assessed the toxicity of sediments containing atrazine singly and as a mixture with the insecticide carbofuran. No effects were observed on *C. tentans* survival when exposed to atrazine singly, and no interactive effects were observed when combined with carbofuran. Fairchild et al. (*61*) investigated the effects of herbicide-insecticide mixtures on the biota of mesocosms. Atrazine did not enhance the bioavailability of esfenvalerate to zooplankton and no ecological synergisms occurred. The author suggests this was due to functional redundancy within the macrophyte community and the dissipation rate of esfenvalerate. The toxicity of atrazine with insecticides was also investigated by Pape-Lindstrom and Lydy (*62*). They showed that atrazine, in combination with the organophosphate insecticides malathion, chlorpyrifos, methyl-parathion and trichlofon produced greater than additive toxic effects to the midge, *C. tentans*. However, atrazine in combination with methoxychlor was determined to be less than additive. The author suggests that the mechanism involved in the greater than additive effects with atrazine and the organophosphates could be related to increased activation of the insecticide through the induction of cytochrome P-450s by atrazine. In a similar study, Jin-Clark et al. (*63*) showed that the mixture of atrazine and chlorpyrifos was greater than additive in toxicity. Although atrazine is not a cholinesterase inhibitor, increased concentrations of atrazine in the mixtures correlated with greater inhibition of cholinesterase. The effects of atrazine were assessed individually and in mixtures with endocrine disrupting compounds (EDCs) (*64*). The 96-hr LC_{50}, 10 day LOEC and NOEC for the estuarian copepod, *E. affinis*, were 125, 49, and 25 μg/L, respectively. Copepods exposed to atrazine at the NOEC showed delayed metamorphosis, while those exposed to atrazine in combination with other EDCs showed impaired development. The author suggests that this might be an additive effect, or likely related to exposure of the EDC in the mixture.

Van den Brink et al. (*65*) studied the chronic effects of atrazine in freshwater microcosms to determine if a 0.1 safety factor multiplied by the lowest available LC or EC_{50} would protect aquatic communities. Their results showed no effects on species composition at an atrazine concentration of 5 μg/L. The benthic communities of experimental ponds treated with atrazine were assessed by Dewey (*66*). Her results showed that non-predatory insects were significantly affected by atrazine at concentrations as low as 20 μg/L. However, predatory insects were not significantly affected. Thus, the effects of atrazine in this study were likely related to reductions in the food supply and loss of habitat of the non-predators. Likewise, deNoyelles et al. (*67*) found that reduced abundance in herbivorous zooplankton was significantly correlated with reduced phytoplankton biomass in experimental ponds treated with either 20 or 500 μg/L atrazine. The direct and indirect effects of environmentally realistic atrazine concentrations (5 µg/L) were assessed at the community level using laboratory microcosms (*68*). No indirect effects due to reduced algal food supply or direct effects related to toxicity were measured. However, a greater number of insects emerged from the treatment mesocosms, suggesting a response to atrazine. A stream sprayed with atrazine during an aerial application to a managed forest showed a significant increase in daytime invertebrate drift immediately following the application (*69*). However, no significant differences were detected in invertebrate densities or number of taxa before and after spraying. The effects of herbicide contaminated biofilm on grazing patterns of invertebrates were studied by Lawrence et al. (*70*). Biofilm contaminated with atrazine and diclofop at 10 and 1µg/L, respectively, had no effect on grazing patterns of invertebrates. Natural benthic assemblages from lake sediments were exposed to atrazine in mesocosms (*71*). At 4 g/L atrazine, shifts in the benthic assemblages were noted, with increased numbers of the gastropod *Fossaria obrussa*, the naiad, *Vejdovskyella intermedia*, and the tubificid worm, *Potamothrix vejdovskyi,* compared to controls. Increases in these organisms could be related to enhanced nutrient recycling and increased numbers of heterotrophic microbial species that would result in increased food availability (*71*). A probabilistic risk assessment for atrazine was conducted by Solomon et al. (*72*). Although invertebrates were not an endpoint in the assessment, risks to this portion of the aquatic ecosystem should be insignificant as well.

Metribuzin

The toxicity of metribuzin to aquatic invertebrates has been assessed using the midge, *Chironomus riparius*, and the water flea, *Ceriodaphnia dubia*. Buhl and Faerber (*73*) tested technical grade metribuzin and the formulated product Sencor® against *C. riparius* and found that the technical material was more toxic than the formulated product. The 48-hr EC_{50} values were 43.5 and 130 mg/L for the technical material and Sencor®, respectively. Ort et al. (*74*) assessed the acute and chronic toxicity of metribuzin formulated as Lexone DF® to *C. dubia*. The 48-hr LC_{50} was determined to be 35.36 mg/L. The NOECs for survival and reproduction after a 7-day exposure were 25 and 6.25 mg/L, respectively.

A risk assessment of metribuzin in aquatic environments was performed by Brock et al. (*75*). They determined that the NOEC for metribuzin to zooplankton was 18 µg/L and that metribuzin at concentrations as high as 180 µg/L had no effect on the macroinvertebrate community. Effects on the zooplankton were believed to be related to changes in the phytoplankton community upon which the zooplankton feed, rather than from direct exposure to the herbicide.

Simazine

A literature review of the toxic effects of simazine on aquatic invertebrates (and other organisms) was performed by Strandberg et al. (*76*). No studies were cited dealing with the effects of simazine in field studies; however, several laboratory studies were described (*77-82*). The most sensitive invertebrate species life-stage tested appeared to be embryos of the snail, *Lymnea stagnalis*, in which an ED_{50} for embryo death was determined to be < 10^{-7} M (0.02 mg/L) (*81*). Another snail species, *Viviparus georgianus*, displayed sensitivity to simazine formulated as Princep 80W® or Aquazine® applied to a lake for blue-green algae control (*83*). At an aqueous concentration of 0.5 mg/L simazine, adults were observed aborting young, all individuals were lethargic and immatures were dying two days after the treatment. However, laboratory exposures at concentrations 10X this amount produced no mortality in *V. georgianus* individuals. Thus, effects observed in the field might be related to an indirect effect of the blue-green algae die off. The interaction of simazine with organophosphate insecticides was studied by Lydy and Austin (*84*). Synergistic ratios of 2.4 and 1.8 were calculated for the binary mixtures of simazine + methidathion and simazine + chlorpyrifos, respectively , indicating greater than additive effects.

Triclopyr

The toxicity of triclopyr formulated as Garlan-3A® to crayfish, *Procambarus* spp., was assessed by Abdelghani et al. (*85*). The reported 48-hr LC_{50} for triclopyr was 28,489.9 mg/L, however, this value is well above the solubility limits for this chemical (440 mg/L). Servizi et al. (*86*) reported a 96-hr EC_{50} of 1.2 mg/L for *Daphnia pulex* exposed to triclopyr ester formulated as Garlon 4®. The toxicity of Garlon 4®, was also tested with field-collected macroinvertebrates from Pacific Northwest streams (*6*). In this study, 96-hr LC_{50} values ranged from 8.1 to 45 mg/L for the stonefly, *Calineuria californica*, and the caddisfly, *Lepidostoma unicolor*, respectively. The hazard concentration to 5% of the theoretical benthic community was determined to be 0.11 mg/L. The length of exposure to Garlon 4® was shown to be directly related to toxicity (*87*). Extending exposure from 9 hr to 24 hr lowered the LC_{50} values for the caddisfly, *Hydropsyche* sp., and the mayfly, *Isonychia* sp., from 14.9 to 4.0 mg/L and 37.0 to 8.8 mg/L, respectively. The route of accumulation (dermal vs. ingestion) was also shown to affect the toxicity of triclopyr ester to aquatic insects (*88*). Their data indicated that organic material (even when ingested) is a sink for triclopyr, making it less bioavailable than its aqueous form.

Garlon-3A® applied to an experimental wetland had no effect on *in situ D. magna* survival, or on sediment dwelling invertebrates (*89*). No significant differences in water column invertebrates collected in traps were observed at 1 day post-spray. However, there was a significant increase in brachiopods and copepods collected 7 days post-spray compared to controls. Kreutzweiser et al. (*90*) assessed acute lethal responses of aquatic insects to triclopyr ester (Garlon 4®) in flow-through laboratory bioassays and lethal and behavioral effects in outdoor stream channels. In the laboratory assays, 10 of the 12 species tested showed no significant increase in mortality at concentrations < 80 mg/L. In outdoor stream channels, triclopyr ester caused significant drift and mortality in the caddisfly, *Dolophilodes distinctus*, at 3.2 mg/L, the stonefly, *Isogenoides* sp., at 32 mg/L and the caddisfly, *Hydropsyche* sp., and mayfly, *Epeorus vitrea*, at 320 mg/L. However, the author suggests that realistic environmental concentrations would not approach concentrations causing observed effects. Garlon 4® applied to a first order stream, which produced instream concentrations of 0.8 and 2.7 μg/L, caused no effects on drift, abundance or richness of the benthic community (*91*). Likewise, Maloney (*92*) demonstrated that triclopyr ester (Grazon®) had no effect on macroinvertebrate richness or abundance in a New Zealand stream after an aerial application that produced a maximum in-stream concentration of 1.5 μg/L.

2.4-D

The toxicity of 2,4-D has been assessed in several invertebrate species. Sanders and Cope (*93*) used the stonefly, *Pteronarcys californica*, reporting a 96-hr LC_{50} of 15 mg/L. Milam et al. (94) compared the acute toxicity of 2,4-D among six species of mussels and two species of water fleas. *Leptodea fragilis* was shown to be the most sensitive mussel, and *Utterbackia imbecillis* was the least sensitive, with 24-hr LC_{50} values of 81.8 and 436.5 mg/L, respectively. *Ceriodaphnia dubia* was shown to be more sensitive than *Daphnia magna,* with 24-hr LC_{50} values of 272.5 and 415.7 mg/L, respectively. The toxicity of 2,4-D in a complex mixture of atmospherically transported pesticides was assessed with the water flea, *C. dubia* (*95*). In this study, 2,4-D was not shown to contribute to the toxicity of the mixture. The MATC for the free acid form of 2,4-D was reported to be <1 mg/L for juvenile Dungeness crab, *Cancer magister* (*96*).

Several studies addressing the toxic effects of 2,4-D to the crab, *Chasmagnathus granulata*, were conducted by Rodriguez and others. Acute toxicity was assessed in larval and juvenile stages of *C. granulata*. The larval stage was shown to be more sensitive than the juvenile stage, with 96-hr LC_{50} values of 0.30 and 2.89 mg/L, respectively (*97*). A 4-week LC_{50} of 30.36 mg/L was also determined for the juvenile stage (*98*). Female *C. granulata* exposed to 2,4-D at a concentration of 15 mg/L showed an increase in morphological abnormalities in hatched larvae compared to controls (*99*). The author suggests that effects observed might have been related to an inhibition of ATP-synthesis.

No direct effects have been observed in macroinvertebrate communities inhabiting experimental mesocosms treated with either 2,4-D BEE or 2,4-D DMA formulations (*22, 100-103*). However, in the study conducted by Stephenson and Mackie (*103*), subtle changes in the macroinvertebrate community were observed 338 days after treatment in that treated ponds demonstrated lower diversity than control ponds. The author suggests that differences in the communities were related to the loss of macrophytes due to the 2,4-D treatment.

Bentazon

Little work has been conducted assessing the effects of bentazon on aquatic invertebrates. Mäenpää et al. (*104*) assessed the bioaccumulation and toxicity of bentazon to the midge, *Chironomus riparius*, and the worm, *Lumbruculus variegates,* in sediments. The 48-hr LC_{50} values for *C. riparius* and *L. variegates* in water were 34.4 and 63.2 mg/L, respectively. Bioaccumulation in *L. variegates* was low and was sediment specific, with bioaccumulation factors ranging from 0.8 to 14.6. Growth of *C. riparius* was significantly reduced at sediment concentrations of 1160 and 4650 mg/kg.

Pendimethalin

Pendimethalin is not very water soluble (0.3 mg/L). Thus, LC_{50} values for aquatic invertebrates have not been widely established, since many invertebrates are not sensitive to this herbicide below water solubility limits (Overmyer, unpublished data). Fliedner (*105*) estimated the EC_{50} value of pendimethalin to *Daphnia magna* to be 448 µg/L. The author also showed that the addition of food to the water increased the toxicity of pendimethalin while the addition of humic acids decreased toxicity. Bioaccumulation of pendimethalin from sediments was determined to be sediment specific, with bioaccumulation factors ranging from 0.1 to 10.3 (*104*).

Isoxaben and Oryzalin

Water collected from highway runoff and spiked with 50 µg/L isoxaben and 200 µg/L oryzalin showed no increase in mortality or decrease in reproduction in 8-d static renewal tests with *Ceriodaphnia dubia* (*106*).

Fungicides

The effect of turf fungicides on aquatic invertebrates has received little attention. Of the nineteen fungicides registered for use on turf in Georgia (*2*), only three have been located in the literature as having been assessed for direct effects on aquatic invertebrates. The most probable reasons for this lack of attention are that the majority of fungicides are not as toxic to aquatic invertebrates, or as widely used, as some of the herbicides and insecticides previously discussed. Thus, lethal or sublethal effects might only be observed at extremely high concentrations which often approach water solubility limits. The three fungicides that have received attention are fenarimol, chlorothalonil, and vinclozolin.

Fenarimol

Fenarimol has been assessed as a potential endocrine disruptor in aquatic invertebrates. A study by Janer et al. (*107*) showed that fenarimol had no effect on testosterone acyltransferase activity in the echinoderm *Paracentrotus lividus,* the gastropod *Marisa cornuarietis*, or the amphipod *Hyalella azteca*. However, sulfation of testosterone was inhibited in *P. lividus* in a dose dependant manner in the range of 1-100μM. In addition, 100μM of fenarimol increased the formation of the testosterone metabolites dihydrotestosterone and 5α-A-diol in *P. lividus* (*108*). Altered testosterone metabolism might have important physiological consequences, and may be a cause of imposex in snails. Fenarimol was also shown to produce antennae and spine abnormalities in developing *Daphnia magna* embryos exposed singly, while exposure to fenarimol and testosterone jointly showed synergistic effects in relation to embryo toxicity (*109*). These data further support the endocrine disrupting effects of fenarimol in invertebrates.

Chlorothalonil

Laboratory and field assessments have been conducted on the effects of chlorothalonil on aquatic invertebrates. Ninety-six hour LC_{50} values of the formulated product Bravo® 500 for the blue mussel, *Mytilus edulis*, and the clam, *Mya arenaria*, were 5.9 mg/L and 35.0 mg/L, respectively (*110*). The 48-h LC_{50} for the water flea, *Daphnia magna*, was between 130 and 200 μg/L. However, concentrations as low as 32 μg/L significantly increased the time to production of first young (*110*). A seven-day LC_{50} of 156μg/L, using the formulated product Dragon Daconil 2787, was reported for the water flea *Ceriodaphnia dubia*, with a seven-day NOEC and LOEC for impaired reproduction of 100 and 125 μg/L, respectively (*111*). Chlorothalonil was shown to be acutely toxic to the freshwater shrimp, *Paratya australiensis*, with a 7-d LC_{50} of 10.9 μg/L (*30*). Glutathione and glutathione-S-transferase activities were also shown to be elevated at chlorothalonil concentrations $\geq$ 0.3 and $\geq$ 1.8 μg/L.

The effects of an aerial application of Bravo® 500 on a pond community were assessed by Ernst and others (*110*). Three applications of Bravo® 500 were applied over a two week period to the experimental pond, at a rate of 875 g a.i./ha. Results showed increased mortality in caged water boatmen, *Sigara alternate*. However, caddisfly larvae, *Limnephilus* sp., freshwater clams, *Psidium* sp., water beetles, *Haliplus* sp., scud, *Gammarus* spp., and midge larvae, Chironomidae, were not affected. The effects in the water boatmen were suspected to be related to their interaction with the surface microlayer, where chlorothalonil concentrations tended to initially concentrate (*112*). Lack of effects observed in the rest of the invertebrate community could be related to factors such as dilution, adsorption to suspended particles and microbial degradation.

Vinclozolin

Vinclozolin has also been assessed for potential endocrine disrupting capabilities in invertebrates. Fresh water snails, *Marisa cornuarietis*, exposed to concentrations of vinclozolin between 0.1 and 1.0 μg/L for three months showed decreased extension of the male accessory sex organs compared to controls (*113*). However, this effect was not apparent after two more months of vinclozolin exposure, when the snails were sexually mature. Similar results were seen with the marine snail, *Nucella lapillus,* exposed to concentrations of vinclozolin between 0.03 and 1.0 μg/L (*113*). Significant reductions in the size of the penis and prostrate gland were observed, as well as a reduced number of males with ripe sperm stored in the seminal vesicles. The author suggests that, although significant effects were obtained in the snails, these effects might not be biologically significant at the population level.

Multiple Pesticides

A few studies reported the effects of exposure to multiple turf pesticides on aquatic invertebrates. Moore et al. (*114*) assessed the toxic effects of contaminated Mississippi oxbow lakes to the amphipod, *Hyalella azteca*. A total of seventeen pesticides were detected in the sediments; five were pesticides that can be applied to turf (atrazine, bifenthrin, fipronil lambda-cyhalothrin, and pendimethalin). Results of their study showed no significant effect of the contaminated sediments on survival or growth of *H. azteca.* Overmyer et al. (*115*) assessed the effects of lawn-care chemical on the macroinvertebrate communities of six suburban streams. Turf chemicals detected in water and/or sediments were chlorothalonil, dithiopyr, oxadiazon, pendimethalin and prodiamine. The results of their study showed a significant correlation between increasing pesticide toxicity in the water and sediment and degradation of the benthic community. However, a majority of the toxicity was attributed to the insecticide chlorpyrifos, which is no longer registered for use on turf. Haith and Rossi (*53*) conducted an ecological risk assessment of pesticide runoff from turf. They determined that the runoff of four pesticides, chlorothalonil, iprodione, PCNB and trichlorfon, might lead to adverse effects on aquatic organisms, including invertebrates.

Conclusion

The literature reviewed in this chapter represents a major portion of the information available concerning the effects of turf pesticides on aquatic invertebrates. While many effects of pesticides were observed in the biota, many of these studies were conducted in the laboratory where study organisms were exposed to high levels of the active ingredients. Several, but not all, studies demonstrated that exposures to environmentally realistic concentrations produced minimal effects in the test organisms. Thus, when applied correctly, many of the pesticides used on turf should produce minimal effects on aquatic invertebrates. However, more field research is needed to determine the effects of pesticides on natural populations, and the indirect effects that might influence overall ecosystem function.

References

1. United States Geological Survey (USGS). USGS Circular 1225: Reston, VA. 1999.
2. *Georgia Pest Management Handbook,Commercial Edition*; Guillebeau, P., Ed.; UGA CAES Special Bulletin 28, Athens, GA 2004.
3. Federle, P. F.; Collins, W. J. *Ohio J. Sci.* **1976**, 76, 19-23.
4. Jadhav, S.; Sontakke, Y. B.; Lomte, V. S. *Environ. Ecol.* **1996**, 14, 863-865.
5. Jadhav, S.; Sontakke, Y. B.; Lomte, V. S. *J. Ecotoxicol. Environ. Monit.* **1996**, 6, 109-112.
6. Peterson J. L.; Jepson, P. C.; Jenkins, J. *J. Environ. Toxicol. Chem.* **2001**, 20, 2205-2214.
7. Conners, D. E.; Black, M. C. *Arch. Environ. Contam. Toxicol.* **2004**, 46, 362-371.
8. Overmyer, J. P.; Armbrust, K. L.; Noblet, R. *Environ. Toxicol. Chem.* **2003**, 22, 1582-1588.
9. Hernández, D. A.; Lombardo, R. J.; Ferrari, L.; Tortorelli, M. C. *Bull. Environ. Contam. Toxicol.* **1990**, 45, 734-741.
10. Hardersen, S.; Wratten, S. D. *New Zeal. J. Mar. Fresh.* **2000**, 34, 117-123.
11. Lohner, T. W.; Fisher, S. W. *Aquat. Toxicol.* **1990**, 16, 335-354.
12. Overmyer, J. P.; Noblet, R. *Arch. Environ. Contam. Toxicol.* **2003**, 24, 209-215.
13. Donkin, P.; Widdows, J.; Evans, S. V.; Staff, F. J.; Yan, T. *Pestic. Sci.* **1997**, 49, 196-209.
14. Bridges, C. M. *Aquat. Ecol.* **1999**, 33, 205-211.
15. Tripathi, P. K.; Singh, A. *Bull. Environ. Contam. Toxicol.* **2002**, 68, 606-611.
16. Seugé, J.; Bluzat, R. *Hydrobiologia* **1983**, 106, 65-72.
17. Krishnan, M.; Chockalingam, S. *Environ. Pollut.* **1989**, 56, 319-326.
18. Hardersen, S.; Frampton, C. M. *Entomol. Exp. Appl.* **1999**, 92, 1-7.
19. Hanazato, T.; Yasuno, M. *Environ. Pollut.* **1987**, 48, 145-159.
20. Hanazato, T.; Yasuno, M. *Arch. Environ. Contam. Toxicol.* **1990**, 19, 77-83.
21. Havens, K. E. *Environ. Pollut.* **1994**, 84, 245-251.
22. Relyea, R. A. *Ecol. Appl.* **2005**, 15, 618-627.
23. Passy, S. I.; Bode, R. W.; Carlson, D. M.; Novak, M. A. *Internat. Rev. Hydrobiol.* **2004**, 89, 121-138.
24 Beyers, D. W.; Farmer, M. S.; Sikoski, P. J. *Arch. Environ. Contam. Toxicol.* **1995**, 28, 27-34.
25. Courtemanch, D. L.; Gibbs, K. E. *Can. Entomol.* **1980**, 112, 271-276.

26. Holmes, S. B.; Millikin, R. L.; Kingsbury, P. D. *Canadian Forestry Service Report FPM-X-46*; Forest Pest Management Institue, Sault Ste. Marie, Ontario, Can.
27. Werner, I.; Deanovic, L. A.; Connor, V.; de Vlaming, V.; Bailey, H. C.; Hinton, D. E. *Environ. Toxicol. Chem.* **2000**, 19, 215-227.
28. Hussain, M. A.; Mohamad, R. B.; Oloffs, P. C. *J. Environ. Sci. Health B.* **1984**, 19, 355-377.
29. Moulton, C. A.; Fleming, W. J., Purnell, C. E. *Environ. Toxicol. Chem.* **1996**, 15, 131-137.
30. Davies, P. E.; Cook, L. S. J.; Goenarso, D. *Environ. Toxicol. Chem.* **1994**, 13, 1341-1354.
31. Rabeni, C. F.; Stanley, J. G. *Bull. Environ. Contam. Toxicol.* **1979**, 23, 327-334.
32. Bocsor, J. G.; O'Connor, T. F. Lake Ontario Environmental Lab; State University College, Oswego, NY. 1975, 174, 29.
33. Overmyer, J. P.; Mason, B. N.; Armbrust, K. L. *Bull. Environ. Contam. Toxicol.* **2005**, 74, 872-879.
34. Ali, A.; Nayar, J. K. Gu, W. D. *J. Am. Mosquito. Contr.* **1998**, 14, 216-218.
35. Chaton, P. F.; Ravanel, P.; Tissut, M.; Meyran, J. C. *Ecotoxicol. Environ. Saf.* **2002**, 52, 8-12.
36. Key, P. B.; Chung, K. W.; Opatkiewicz, A. D., Wirth, E. F.; Fulton, M. H. *Bull. Environ. Contam. Toxicol.* **2003**, 70, 533-540.
37. United States Environmental Protection Agency (US EPA) *New Pesticide Fact Sheet, EPA 737-F-96-005*; US EPA Washington DC. 1996.
38. Schlenk D.; Huggett, D. B.; Allgood, J.; Bennett, E.; Rimoldi, J.; Beeler, A. B.; Block, D.; Holder, A. W.; Hovinga, R.; Bedient, P. *Arch. Environ. Contam. Toxicol.* **2001**, 41, 325-332.
39. Konwick, B. J.; Fisk, A. T.; Garrison, A. W.; Avants, J. K.; Black, M. C. *Environ. Toxicol. Chem.* **2005**, 24, 2350-2355.
40. Chandler, G. T.; Cary, T. L.; Volz, D. C.; Walse, S. S.; Ferry, J. L.; Klosterhaus, S. L. *Environ. Toxicol. Chem.* **2004**, 23, 117-124.
41. Chandler, G. T.; Cary, T. L.; Bejarano, A. C.; Pender, J.; Ferry, J. L. *Environ. Sci. Technol.* **2004**, 38, 6407-6414.
42. Cary, T. L.; Chandler, G. T.; Volz, D. C.; Walse, S. S.; Ferry, J. L. *Environ. Sci. Technol.* **2004**, 38, 522-528.
43. Stark, J. D.; Vargas, R. I. *Ecotoxicol. Environ. Saf.* **2005**, 62, 11-16.
44. Volz, D. C.; Wirth, E. F.; Fulton, M. H.; Scott, G. I.; Strozier, D. S.; Block, D. S.; Ferry, J. L.; Walse, S. S.; Chandler, G. T. *Bull Environ. Contam. Toxicol.* **2003**, 71, 497-503.
45. Song, M. Y.; Stark, J. D.; Brown, J. J. *Environ. Toxicol. Chem.* **1997**, 16, 2494-2500.
46. Mokry, L. E.; Hoagland, K. D. *Environ. Toxicol. Chem.* **1990**, 9, 1045-1051.
47. Liu, W.; Gan, J. J.; Qin, S. *Chirality* 2005, 17, S127-S133.
48. Amweg, E. L.; Weston, D. P.; Ureda, N. M. *Environ. Toxicol. Chem.* **2005**, 24, 966-972.
49. Heimbach, F.; Pflueger, W.; Ratte, H. T. *Envrion. Toxicol. Chem.* **1992**, 11, 27-34.

50. Weston, D. P.; You, J.; Lydy, M. J. *Environ. Sci. Technol.* **2004**, 38, 2752-2759.
51. Heckman, L. H.; Friberg, N. *Environ. Toxicol. Chem.* **2005**, 24, 582-590.
52. Solomon, K, R.; Giddings, J. M.; Maund, S. J. *Environ. Toxicol. Chem.* **2001**, 20, 652-659.
53. Haith, D. A.; Rossi, F. S. *J. Environ. Qual.* **2003**, 32, 447-455.
54. Macek, K. J.; Buxton, K. S.; Sauter, S.; Gnilka, S.; Dean, J. W. *Ecol. Res. Ser. EPA-600/3-76-047*, US EPA, Washington DC, 1976.
55. Plumley, F. G.; Davis, D. E.; McEnerney, J. T.; Everest, J. W. *Estuaries.* **1980**, 3, 217-223.
56. Taylor, E. J.; Maund, S. J.; Pascoe, D. *Arch. Environ. Contam. Toxicol.* **1991**, 21, 371-376.
57. Streit, B.; Peter, H. M. *Arch. Hydrobiol./Suppl55.* **1978**, 1, 62-77.
58. Hall, L. W.; Ziegenfuss, M. C.; Anderson, R. D.; Tierney, D. P. *Arch. Envion. Contam. Toxicol.* **1995**, 28, 344-348.
59. Phyu, Y. L.; Warne, M. St. J.; Lim, R. P. *Chemosphere* **2005**, 58, 1231-1239.
60. Douglas, W. S.; McIntosh, A. Clausen, J. C. *Environ. Toxicol. Chem.* **1993**, 12, 847-853.
61. Fairchild, J. F.; LaPoint, T. W.; Schwartz, T. R. *Arch. Environ. Contam. Toxicol.* **1994**, 27, 527-533.
62. Pape-Lindstrom, P. A.; Lydy, M. J. *Environ. Toxicol. Chem.* **1997**, 16, 2415-2420.
63. Jin-Clark, Y.; Lydy, M. J.; Zhu, K. Y. *Environ. Toxicol. Chem.* **2002**, 21, 598-603.
64. Forget-Leray, J.; Landriau, I.; Minier, C.; Leboulenger, F. *Ecotoxicol. Environ. Saf.* **2005**, 60, 288-294.
65. van den Brink, P. J.; van Donk, E.; Gylstra, R.; Crum, S. J. H.; Brock, T. C. M. *Chemosphere* **1995**, 31, 3181-3200.
66. Dewey, S. L. *Ecology* **1986**, 67, 148-162.
67. deNoyelles, F.; Kettle, W. D.; Sinn, D. E. *Ecology*. **1982**, 63, 1285-1293.
68. Gruessner, B.; Watzin, M. C. *Environ. Toxicol. Chem.* **1996**, 15, 410-419.
69. Davies, P. E.; Cook, L. S. J.; Barton, J. L. *Aust. J. Mar. Feshwater Res.* **1994**, 45, 209-26.
70. Lawrence, J. R.; Scharf, B.; Packroff, G.; Neu, T. R. *Microb. Ecol.* **2002**, 43, 199-207.
71. Fletcher, R; Reynoldson, T. B.; Taylor, W. D. Environ. Pollut. **2001**, 115, 173-182.
72. Solomon, K. R.; Baker, D. B.; Richards, R. P.; Dixon, K. R.; Klaine, S. J.; LaPoint, T. W.; Kendall, R. J.; Weisskopf, C. P.; Giddings, J. M.; Giesy, J. P.; Hall, L. W.; Williams, W. M. *Environ. Toxicol. Chem.* **1996**, 15, 31-76.
73. Buhl, K. J.; Faerber, N. L. *Arch. Environ. Contam. Toxicol.* **1989**, 18, 530-536.
74. Ort, M. P.; Fairchild, J. F.; Finger, S. E. *Arch. Environ. Contam. Toxicol.* **1994**, 27, 103-106.
75. Brock, T. C. M.; Crum, S. J. H.; Deneer, J. W.; Heimbach, F.; Roijackers, R. M. M.; Sinkeldam, J. A. *Environ. Pollut.* **2004**, 130, 403-426.

76. Strandberg, M. T.; Scott-Fordsmand, J. *J. Sci. Total Environ.* **2002**, 296, 117-137.
77. Sanders, H. O. *Tech. Paper Bur. Sport Fish Wildl.* **1969**, 25, 1-17.
78. Sanders, H. O. *J. Water Pollut. Control Fed.* **1970**, 42, 1544-1550.
79. Fritmayer, K. M.; Geiger, J. G.; Van-Den-Avyle, M. J. *Arch. Environ. Contam. Toxicol.* **1982**, 11, 603-609.
80. Foster, S.; Thomas, M.; Korth, W. Austral.. *J. Ecotoxicol.* **1998**, 4, 53-59.
81. Kosanke G. J.; Schwippert, W. W.; Beneke, T. W. *Comp. Biochem. Physiol. C Comp. Pharmacol. Toxicol.* **1988**, 90, 373-380.
82. Sakar, S. K. *Geobios* **1997**, 24, 142-145.
83. Harman, W. N. *The Nautilus.* **1978**, 92, 129-134.
84. Lydy, M. J.; Austin, K. R. *Arch. Environ. Contam. Toxicol.* **2004**, 48, 49-55.
85. Abdelghani, A. A.; Tchounwou, P. B.; Anderson, A. C.; Sujono, H.; Heyer, L. R.; Monkiedje, A. *Environ. Toxicol. Water Qual.* **1997**, 12, 237-243.
86. Servizi, J. A.; Gordon, R. W.; Martens, D. W. *Bull. Environ. Contam. Toxicol.* **1987**, 39, 15-22.
87. Kreutzweiser, D. P.; Holmes, S. B.; Eichenberg, D. C. *Arch. Environ. Contam. Toxicol.* **1994**, 26, 124-129.
88. Kreutzweiser, D. P.; Thompson, D. G.; Stanzik, B.; Shepherd, J. A. *J. Environ. Qual.* **1998**, 27, 1138-1147.
89. Gardner, S. C.; Grue, C. E. *Environ. Toxicol. Chem.* **1996**, 15, 441-451.
90. Kreutzweiser, D. P.; Holmes, S. B.; Behmer, D. J. *Ecotoxicol. Environ. Saf.* **1992**, 23, 264-374.
91. Thompson, D. G.; Kreutzweiser, D. P.; Capell, S. S. *Environ. Toxicol. Chem.* **1995**, 14, 1307-1317.
92. Maloney, R. F. *New Zeal. J. Mar. Fresh.* **1995**, 29, 505-515.
93. Sanders, H. O.; Cope, O. B. *Limnol. Oceanog.* **1968**, 13, 112-117.
94. Milam, C. D.; Farris, J. L.; Dwyer, F. J.; Hardesty, D. K. *Arch. Environ. Contam. Toxicol.* **2005**, 48, 166-173.
95. George, T. K.; Waite, D.; Liber, K.; Sproull, J. *Environ. Monit. Assess.* **2003**, 85, 309-326.
96. Caldwell, R. S.; Buchanan, D. V.; Armstrong, D. A.; Mallon, M. H.; Millemann, R. E. *Arch. Environ. Contam. Toxicol.* **1979**, 8, 383-396.
97. Rodriguez, E. M.; Amin, O. A. *Bull. Environ. Contam. Toxicol.* **1991**, 47, 634-640.
98. Rodriguez, E. M.; Monserrat, J. M.; Amin, O. A. *Arch. Environ. Contam. Toxicol.* **1992**, 22, 140-145.
99. Rodriguez, E. M.; Pisano, A. *Comp. Biochem. Physiol. C.* **1993**, 104, 71-78.
100. Pierce, M. E. *Proc. Northeast Weed Control. Conf;* **1960**, 14, 483-487.
101. Smith, G. E.; Isom, B. G.; *Pestic. Monit. J.* **1967**, 1, 16-21.
102. Boyle, T. P. *Environ. Pollut.* **1980**, A21, 35-49.
103. Stephenson, M.; Mackie, G. L. *Aquat. Toxicol.* 1986, 243-251.
104. Mäenpää, K. A.; Sormunen, A. J.; Kukkonen, J. V. K. *Ecotoxicol. Environ. Saf.* **2003**, 56, 398-410.
105. Fliedner, A. *Chemosphere,* **1997**, 35, 295-305.
106. Huang, X.; Fong, S.; Deanovic, L.; Young, T. M. *Environ. Toxicol. Chem.* **2005**, 24, 2336-2340.

107. Janer, G.; Sternberg, R. M.; LeBlanc, G. A.; Porte C. *Aquat. Toxicol.* **2005**, 71, 273-282.

108. Janer, G.; LeBlanc, G. A.; Porte, C. *Gen. Comp. Endocr.* **2005**, 143, 211-221.

109. Mu, X.; LeBlanc, G. A. *J. Exp. Zool.* **2002**, 292, 287-292.

110. Ernst, W.; Doe, K.; Jonah, P.; Young, J.; Julien, G.; Hennigar, P. *Arch. Environ. Contam. Toxicol.* **1991**, 21, 1-9.

111. Sherrard, R. M.; Murray-Gulde, C. L.; Rodgers, J. H.; Shah, Y. T. *Ecotoxicol. Environ. Saf.* **2003**, 56, 327-333.

112. Muir, D. C. G.; Rawn, G. P.; Grift, N. P. *J. Agric. Food Chem.* **1985**, 33, 603-609.

113. Tillman, M.; Schulte-Oehlmann, U.; Duft, M.; Markert, B.; Oehlmann, J. *Ecotoxicol.* **2001**, 10, 373-388.

114. Moore, M. T.; Lizotte, R. E.; Cooper, C. M.; Smith, S.; Knight, S. S. Bull. *Environ. Contam. Toxicol.* **2004**, 72, 777-783.

115. Overmyer, J. P.; Noblet, R.; Armbrust, K. L. *Environ. Pollut.* **2005**, 137, 263-272.

Chapter 4

Determination of Transferable Residues of Carbaryl from Turf

Michael E. Krolski,[1] Joseph R. Hudson,[2] Vicky Standart,[1] Curt Lunchick,[2] and James K. Campbell[1]

[1]**Bayer CropScience, 17745 S. Metcalf Ave., Stilwell, KS 66085**
[2]**Bayer CropScience, 2 T. W. Alexander Drive, Research Triangle Park, NC 27709**

Bayer CropScience performed a study to determine the specific transferable residues of carbaryl from turf following one application of Sevin® 2G at the maximum proposed label rate of 0.18 lb ai / 1000 square feet. Trials were performed on established turf stands in Florida (St. Augustine grass), Kansas (fescue), and California (bermuda grass) where cloth dosimeter samples were collected, using the Modified California Roller Technique, from both irrigated and non-irrigated plots following application of Sevin® 2G. Additionally, the transferability of residue from turf to hands was measured in the Kansas test using dry and moist hand wipes. Transferable carbaryl residues on cloth dosimeters were initially much lower for irrigated plots; however, by 72 hours post-treatment, the transferable residues from both non-irrigated and irrigated plots were comparable. Similarly, initial carbaryl residues from both dry and moist hand wipes were lower in samples from the irrigated plot compared to similar samples collected from the non-irrigated plot, and the transferable residues declined rapidly with time.

Introduction

The United States Environmental Protection Agency (USEPA) requires data on Transferable Turf Residues (TTR) and Transfer Coefficients (TC) on pesticides applied to residential turf. The data are used in assessing post-application exposure to individuals from treated lawns.

A TTR study was performed for carbaryl used on residential turf. The study was designed based on the requirements of USEPA Series 875: Occupational and Residential Exposure Test Guidelines (*1*) in accordance with USEPA FIFRA Good Laboratory Practices Standards, 40 CFR 160 (*2, 3*). The objective of this study was to characterize the decline of Sevin® 2G transferable residues when applied to turf.

Experimental

Test Substance

Carbaryl (1-naphthyl N-methylcarbamate, CAS Registry No. 63-25-2) is the active ingredient (ai) in Sevin 2G. Sevin 2G, the test substance used in this experiment, is a granular formulation containing 2% by weight of carbaryl. The test substance used for this study came from two different batches and was obtained at a home improvement store in Kansas City, MO. The material was analyzed for active ingredient loading and then was shipped to the field test sites where it was stored at ambient conditions prior to use.

The formulated material used in this study was analyzed by Bayer CropScience (BCS) prior to use. The measured concentrations found were 1.91% and 2.11% carbaryl (wt/wt), respectively. These values are within the specification of the commercial formulation, therefore the nominal value of 2.0% (wt/wt) was used for calculations.

Locations

Field trials were conducted at three sites: Molino, FL (St. Augustine grass, USEPA Region 3), Stilwell, KS (fescue grass, USEPA Region 5), and Fresno, CA (bermuda grass, USEPA Region 10). In all three of the field trials, cloth dosimeter samples (5690 cm^2 total area) were collected from triplicate irrigated and non-irrigated turf subplots immediately after application, and at target times of 4-, 10-, 24-, 48-, 72-, 120-, and 168-hours following the application. In the trial conducted at Stilwell, additional wet and dry hand press samples (343 cm^2 total area) were collected from the irrigated and non-irrigated turf plots immediately after application and 120-hours later. All samples were immediately frozen and transferred to the Bayer Research Park (BRP) for frozen storage prior to analysis.

Experimental Design

The three field trials were conducted on established turf mowed to a height of approximately 2 in (5 cm) at least 48 hours prior to application; grass clippings were removed from the plots at the time of the last mowing. Two plots were included in each trial: one was designated as the non-irrigated plot, the other plot was designated as the irrigated plot. Each plot was divided into three subplots from which the three replicate samples were collected at each time point. Each plot was treated with Sevin 2G at a target application rate of 0.18 lb ai/1000 ft^2 (8.8 kg ai/ha). Immediately following application, one plot in each trial was irrigated with 0.25 in to 0.50 in of water using a lateral water gun. Agronomic practices typical of the trial locations were used for growing and maintaining the turf. During the course of sample collection, the plots were not mowed. The trial site conditions, including soil characteristics, are listed in Table I.

Sample Handling and Preparation

Cloth Dosimeters

In all three of the trials, transferable carbaryl residues were sampled from the three subplots of both the irrigated and non-irrigated plots using the Modified California Roller technique (*4, 5*). The Modified California Roller technique used for measuring the amount of transferable turf residues was developed by the ORETF (Outdoor Residential Exposure Task Force) as a standardized sampling technique. It has been accepted by USEPA, the California Department of Pesticide Regulation and Health Canada, and has been used in many transferable turf residue studies.

Samples were collected pre-application (control samples), immediately post application (0-time), and at 4-, 10-, 24-, 48-, 72-, 120-, and 168-hours after treatment from the non-irrigated plots. Samples were collected pre-application (control samples), as soon as the turf surface was dry following post-application irrigation (0-time), and 4-, 10-, 24-, 48-, 72-, 120-, and 168-hours after treatment from the irrigated plots.

Samples were collected from each subplot at each time point using a white 100% cotton percale 200-thread cloth (27 in x 39 in) inside a rigid frame, with an available exposed surface area of 24.5 in by 36 in (5,690 cm^2), backed by a heavy gauge plastic sheet. The frame was then anchored to the treated subplot, with the exposed cloth in contact with the turf, using 16 penny nails. A roller with a mass of one slug (32 lbs) was then passed back and forth over the plastic backing five times.

Table I. Trial Site Conditions for Carbaryl on Turf

	Soil Characteristics[a]				*Meteorological Data*[b]	
Study Location (City, State)	*Type*	*% OM*	*pH*	*CEC*	*Total Rainfall (in)*[c]	*Temp. Range (°F)*
Molino, FL	Sandy Loam	2.2	6.5	2.0	2.48	50 – 79
Stilwell, KS	Silty Clay Loam	3.1	6.6	13.4	3.46	56 – 90
Fresno, CA	Sandy Loam	0.52	7.3	3.8	0.0	63 - 102

[a] % OM = Percent Organic Matter; CEC = Cation Exchange Capacity

[b] Data is for the interval from first application to last sampling.

[c] Irrigation values for the irrigated plot were not included in the total. Rainfall at the Molino, FL site occurred between the 120-hour and 168-hour sampling events. Rainfall at the Stilwell, KS site occurred between the 24-hour and 72-hour sampling events and precluded collection of the 48-hour samples.

Once each cloth had been rolled, it was shaken to remove any pieces of grass or other debris and was folded with the treated area facing inward (not exposed to the container). The folded cloth was placed into two nested Ziploc® bags, with a sample label affixed to the inner bag. The labeled sample bags were placed either into a cooler with dry ice for temporary storage or directly transferred into a freezer.

All samples were stored frozen within 4 hours after collection, and remained in frozen storage until analyzed.

Hand Presses

Surface residues of pesticides on turf may be transferred to hands, which can then be inserted into the mouth. This circumstance has been cited by EPA as presenting a potential for exposure to children. Young children may engage in hand-to-mouth activities by sucking on their fingers, and they may have wet hands. The trial conducted at Stilwell, KS included the collection of both dry and moist hand press samples from the irrigated and non-irrigated plots immediately after treatment and 120 hours after application to determine potential transfer of residues via hand contact with treated turf.

Ten volunteers were used to collect hand press samples from each treated plot at each time point. All volunteers completed consent forms and the study design was approved by an independent Institutional Review Board charged with insuring the ethical use of human subjects in research. One set of hand presses was dry and the other set was water-moistened with approximately 0.3 mL of water applied from a spray bottle immediately prior to sampling. Each bare hand was placed in contact with the turf seven (7) times with a pressure of approximately 17 lbs (equivalent to 2.1 psi) for approximately 6 sec each time.

Each of the seven hand presses was done in a new location on the turf plot, placing a template with seven square cutouts (7 cm x 7 cm) over the treated turf, so that undisturbed treated turf was contacted each time. The hand press contact area was 49 cm^2, and the cumulative turf contact area for the seven hand presses was 343 cm^2 [(49 cm^2 /press) x 7 presses].

Each of the 10 volunteers thoroughly washed his/her hands with soap (without skin lotion additives) and water followed by thorough towel and air-drying immediately prior to collecting the hand press samples.

Once the hand press was complete, any debris or granules adhering to the hand was carefully removed using tweezers or forceps. The test palm was then wiped twice in succession, each time with one package of two gauze pads (4 in x 4 in sterile cotton pad) moistened with 5 ml of 0.01% Aerosol OT (AOT, aqueous sodium dioctyl sulfosuccinate and butyl cellosolve) solution. The four gauze pads were combined as one sample.

Each gauze pad sample, consisting of 4 pads, was placed into a pre-labeled amber glass jar with a teflon-coated lid. The lid was closed and sealed with electrical tape and the sealed jar was immediately placed in a cooler with dry ice for temporary storage. All samples were placed in a freezer within 4 hours after collection, and remained in frozen storage until analyzed.

Control samples were collected from the plots to be treated prior to application of the test substance.

Field Fortification Samples

Each trial included cloth dosimeter field fortification samples (1 control, 3 low level, and 3 high level) to insure the integrity of the samples during storage and shipment. Cloth dosimeter samples were amended with either 5 µg or 500 µg of carbaryl at each test site. Fortification was done by pipetting 2.0 mL of an acetonitrile solution containing the fortification standard onto a control dosimeter and allowing the solvent to evaporate. These samples were handled, stored and shipped under the same conditions as the treated field samples.

The trial conducted at Stilwell, KS included hand press field fortification samples (1 control, 3 low level, and 3 high level) to insure the integrity of the samples during storage and shipment. Gauze wipes were amended with either 5 µg or 50 µg of carbaryl. Fortification was done by pipetting 2.0 mL of an acetonitrile solution containing the fortification standard onto control gauze pads and allowing the solvent to evaporate. These samples were handled, stored and shipped under the same conditions as the treated field samples.

Sample Analysis

Cloth dosimeter and hand wipe samples were analyzed at ABC Labs in Columbia, MO. The method for determining carbaryl residue on cloth dosimeters was validated by measuring the carbaryl residue recoveries from control samples fortified in the laboratory with 1 µg, 5 µg, and 50 µg carbaryl using 100% acetonitrile as the extraction solvent. The method for determining carbaryl residue on cloth dosimeters was additionally validated by measuring the carbaryl residue recoveries from control samples fortified in the laboratory with 1 μg, 50 μg, 500 μg, and 6500 μg carbaryl using 10% methanol/90% acetonitrile as the extraction solvent. The method for determining carbaryl residue on hand wipes was validated by measuring the residue recoveries from control samples fortified in the laboratory with 0.1 μg, 5 μg, and 50 μg carbaryl.

Concurrent recoveries of carbaryl residue on cloth dosimeters and hand wipes from hand presses were measured during sample analysis to verify method performance. Concurrent recoveries were performed from control cloth dosimeter samples fortified in the laboratory with 1 μg, 5 μg, 50 μg, and 500 μg carbaryl. Concurrent recoveries were performed from control hand wipes fortified with 0.1 μg, 5 μg, and 50 μg carbaryl.

Measured residues on both cloth dosimeters and hand wipes from hand presses were adjusted using recoveries from the control samples fortified in the field with known amounts of carbaryl.

Cloth Dosimeters

Carbaryl was extracted from cloth dosimeters by shaking the dosimeter in a gallon glass jar with 1000 mL of acetonitrile or, in later extraction sets, 10% methanol/90% acetonitrile. Following 30 minutes of shaking, a 250-mL aliquot was removed and transferred to a flat-bottom flask. The extraction solvent was evaporated to dryness by rotary evaporation using a warm water bath at 35 to 40°C. The residue was reconstituted in 60% methanol/40% water containing 2 mL/L 10% phosphoric acid. Carbaryl was quantitated by high-performance liquid chromatography (HPLC) analysis with post-column hydrolysis and fluorescence detection. Laboratory fortifications (controls fortified prior to adding the extraction solvent) were extracted with each set of treated samples. At least one field fortification sample was extracted with each analytical set to allow correction for field recovery.

Hand Presses/Hand Wipes

Carbaryl was extracted from hand wipe samples by shaking twice with 100 mL of acetone. The acetone extracts were combined after passing through sodium sulfate supported on a powder funnel. Following the second extraction, the extraction vessels were rinsed with acetone (15 mL) and the rinse was added to the combined extract. The acetone extract was evaporated to dryness using rotary evaporation with a water bath at 30 to 35°C. The final extract was reconstituted in methanol with sonication and swirling. Sufficient water was then added to the methanolic solution to yield a solution that was 60% methanol and 40% water. Carbaryl was quantitated by HPLC analysis with post-column hydrolysis and fluorescence detection. Laboratory fortifications (controls fortified prior to adding the extraction solvent) were extracted with each set of treated samples. For moist hand wipes, control hand wipes were wetted with 0.01 % Aerosol OT-75 solution prior to fortification and extraction. For dry hand wipes, control hand wipes were fortified and extracted. At least one field fortification sample was extracted with each analytical set to allow correction for field recovery.

Calculations

Total carbaryl residue values on cloth dosimeters and hand wipes were calculated by comparison of instrumental response to a standard curve. The resulting values represented a gross "measured residue" value, expressed in μg, for each analytical sample. The measured residue value was then corrected by dividing by the measured residue determined for the corresponding field fortification recovery value for that set. If more than one field fortification sample was analyzed with a set, the field fortification recovery closest to the measured value was used for the correction. Finally, the residue per unit area was calculated by dividing the corrected residue value by the sampled area (5690 cm^2 for cloth dosimeters or 49 cm^2 for the exposed hand area in the hand press samples).

Cloth Dosimeter Samples

Example calculations for a typical cloth dosimeter sample are shown below:

Measured residue = 2135 μg

Field fortification recovery (500 μg field fortification) = 97%

Cloth dosimeter area = 5690 cm^2

Corrected residue = (2135 μg)/0.97 = 2201 μg

Residue per unit area = (2201 μg)/(5690 cm^2) = 0.38682 μg/cm^2

Average values were then determined for each plot/time-point combination from the three individual subplot samples. Finally, corresponding plot/time-point values were averaged across the three trials that were performed.

Hand Wipe Samples

Example calculations for a typical hand wipe sample are shown below:

Measured residue = 7.40 μg

Field fortification recovery (5 μg field fortification) = 100%

Hand press area = 49 cm^2

Corrected residue = (7.40 μg)/1.00 = 7.40 μg

Residue per unit area = (7.40 μg)/(49 cm^2) = 0.15102 μg/cm^2

Average values from the ten individual replicate samples were then determined for each plot/wet-dry/time-point combination.

Results and Discussion

The relative response of the detector in the chromatographic system to carbaryl was linear over the range of 0.010 μg/mL to 0.500 μg/mL. The correlation coefficients were >0.999 for both cloth dosimeter samples and hand wipe samples. Control interferences ranged from less than the Limit of Quantitation (<LOQ, 1 μg), corresponding to 0.00018 μg/cm^2; to 1.86 μg, corresponding to 0.00033 μg/cm^2 for cloth dosimeters. Control interferences ranged from <LOQ (0.1 μg), corresponding to 0.00204 μg/cm^2; to 0.382 μg, corresponding to 0.00780 μg/cm^2 for hand wipes from hand presses.

Method Validation, Field Fortification and Laboratory Validation Recoveries

Method validation recoveries of carbaryl residue from cloth dosimeters extracted using 100% acetonitrile and fortified at 1 μg (0.00018 μg/cm^2), 5 μg (0.00088 μg/cm^2), or 50 μg (0.0088 μg/cm^2) ranged from 76% to 86% (mean of 82% ± 3.42), 72% to 85% (mean of 80% ± 4.45), and from 78% to 84% (mean of 81% ± 2.27), respectively. Method validation recoveries of carbaryl residue from cloth dosimeters extracted using 9:1 acetonitrile/methanol, fortified at 1 μg (0.00018 μg/cm^2), 50 μg (0.0088 μg/cm^2), 500 μg (0.088 μg/cm^2), or 6500 μg (1.142 μg/cm^2) ranged from 82% to 92% (mean of 87% ± 3.25), 86% to 90% (mean of 87% ± 2.31), 76% to 93% (mean of 86% ± 8.89), and from 86% to 108% (mean of 98% ± 7.04), respectively. Method validation recoveries of carbaryl residue from hand wipes fortified at 0.1 μg (0.00204 μg/cm^2), 5 μg (0.102 μg/cm^2), or 50 μg (1.02 μg/cm^2) ranged from 83% to 89% (mean of 86% ± 2.19), 85% to 89% (mean of 87% ± 1.72), and from 95% to 103% (mean of 98% ± 2.70), respectively.

Field fortification recoveries of carbaryl residue from cloth dosimeters extracted using 100% acetonitrile and fortified at 5 μg (0.00088 μg/cm^2) or 500 μg (0.088 μg/cm^2) ranged from 41% to 42% (mean of 42% ± 0.7), and were 46%, respectively. Field fortification recoveries of carbaryl residue from cloth dosimeters extracted using 9:1 acetonitrile/methanol, fortified at 5 μg (0.00088 μg/cm^2) or 500 μg (0.088 μg/cm^2) ranged from 40% to 103% (mean of 79% ± 17.7) and from 59% to 97% (mean of 82% ± 14), respectively. Field fortification recoveries of carbaryl residue from hand wipes fortified at 5 μg (0.102 μg/cm^2) or 50 μg (1.02 μg/cm^2) ranged from 98% to 104% (mean of 101% ± 2.6) and 73% to 102% (mean of 88% ± 12.6), respectively.

Laboratory fortification recoveries of carbaryl residue from cloth dosimeters extracted using 100% acetonitrile fortified at 1 μg (0.00018 μg/cm^2) or 50 μg (0.0088 μg/cm^2) were 78% and 72%, respectively. Lab fortification recoveries of carbaryl residue from cloth dosimeters extracted using 9:1 acetonitrile/methanol fortified at 1 μg (0.00018 μg/cm^2) or 50 μg (0.0088 μg/cm^2) ranged from 73% to 105% (mean of 83% ± 10.4) and from 78% to 98% (mean of 91% ± 6.3), respectively. Lab fortification recoveries of carbaryl residue from dry hand wipes fortified at 0.1 μg (0.00204 μg/cm^2), 5 μg (0.102 μg/cm^2), or 50 μg (1.02 μg/cm^2) ranged from 94% to 107% (mean of 101% ± 9.2), 89% to 97% (mean of 93% ± 5.7), and from 93% to 98% (mean of 96% ± 3.5), respectively. Lab fortification recoveries of carbaryl residue from moist hand wipes fortified at 0.1 μg (0.00204 μg/cm^2), 5 μg (0.0102 μg/cm^2), or 50 μg (1.02 μg/cm^2) ranged from 81% to 95% (mean of 88% ± 9.9), 83% to 97% (mean of 90% ± 9.9), and from 81% to 90% (mean of 86% ± 6.4), respectively.

All recoveries were corrected for any interferences in corresponding controls.

Limit of Quantitation (LOQ) and Limit of Detection (LOD)

The LOQ is defined as the lowest fortification level of an analyte at which acceptable recovery can be achieved. Acceptable recovery of carbaryl residue from cloth dosimeters was achieved at 1 μg (0.00018 μg/cm^2), therefore, the LOQ for carbaryl residue in cloth dosimeters was 1 μg (0.00018 μg/cm^2). Acceptable recovery of carbaryl residue from hand wipes was achieved at 0.1 μg (0.00204 μg/cm^2), therefore, the LOQ for carbaryl residue in hand wipes was 0.1 μg (0.00204 μg/cm^2). Any residue found to be less than the LOQ was reported as less than the LOQ, regardless of the value.

The LOD is defined as the lowest concentration of an analyte that can be determined to be statistically different from a blank. The LOD was determined from method validation and concurrent recovery data obtained from control samples fortified with carbaryl at the LOQ. The LOD was calculated by multiplying the standard deviation of recovery measurements at the LOQ by $t_{0.99}$ (where $t_{0.99}$ is the one-tailed t-statistic at the 99% confidence level for n-1 replicates) (*6*) and rounding up to the next higher thousandth of a ppm. The LOD values for carbaryl residues in cloth dosimeters and hand wipes were 0.106 μg (0.00002 μg/cm^2) and 0.007 μg (0.00014 μg/cm^2), respectively. Any measured residue that was less than the LOD was reported as less than the LOD, regardless of the value.

Magnitude of Carbaryl Residue on Cloth Dosimeters and Hand Wipes from Hand Presses

Cloth Dosimeters

Average carbaryl residue data on cloth dosimeters from non-irrigated and irrigated plots are given in Table II and shown graphically in Figure 1. In both plots, the highest transferable carbaryl residue was observed immediately following application. The residue rapidly declined over the first 4 hours and then leveled off or rose slightly after 10 hours. The slight increase in transferable residue at the 10-hour time point could have been the result of turf dampness from the evening dew. At later time points, transferable residue was very low.

Transferable residues were initially much lower for irrigated plots, averaging >65% lower (0.126 $\mu g/cm^2$) than the corresponding non-irrigated plots (0.388 $\mu g/cm^2$). However, by 72 hours post-treatment, the transferable residues from both non-irrigated and irrigated plots were comparably low.

Transfer to cloth dosimeters at the first collection time point ranged from 0.046% to 0.622% of the theoretical amount of residue available, with an average of 0.142% for irrigated plots and 0.441% for non-irrigated plots. By 168 hours post-treatment, transfer to cloth dosimeters corresponded to <0.003% of the theoretical amount of residue, in all plots.

Table II. Carbaryl Residue from Cloth Dosimeters Collected from Turf Treated with Sevin® 2G

Hours Post-Treatment	*Residue ($\mu g/cm^2$)* Pensacola FL	Stilwell KS	Fresno CA	Average
Non-Irrigated Plots				
0	0.548	0.166	0.450	0.388
4	0.059	0.062	0.025	0.049
10	0.267	0.050	0.014	0.110
24	0.073	0.009	0.047	0.043
48	0.082	NC[a]	0.016	0.049
72	0.019	0.000	0.003	0.008
120	0.001	0.001	0.003	0.002
168	0.002	0.000	0.002	0.001
Irrigated Plots				
0	0.288	0.048	0.040	0.125
4	0.196	0.006	0.008	0.070
10	0.040	0.004	0.003	0.015
24	0.011	0.004	0.025	0.013
48	0.019	NC[a]	0.017	0.018
72	0.020	0.001	0.010	0.010
120	0.002	0.000	0.004	0.002
168	0.001	0.001	0.001	0.001

[a] Sample was not collected due to rain.

Hand Wipes from Hand Presses

Carbaryl residue data for hand wipes from hand presses at both time points are given in Table III and shown graphically in Figure 2. Transferable carbaryl residues in the non-irrigated plot, averaging 0.242 µg/cm^2 and 0.583 µg/cm^2, were observed immediately following application for dry and moist hand wipes, respectively. Transferable carbaryl residues in the irrigated plot, averaging 0.064 µg/cm^2 and 0.083 µg/cm^2, were observed immediately following application for dry and moist hand wipes, respectively. By the 120-hour time point, residues had declined to <LOQ (0.0020 µg/cm^2) in all plots.

Transfer of carbaryl residues to hand presses at the first collection from the irrigated plot ranged from 0.07% to 0.09% of the theoretical amount of residue available for dry and moist hand wipes, respectively (Table IV). Transfer of carbaryl residues to hand presses at the first collection from the non-irrigated plot ranged from 0.27% to 0.66% of the theoretical amount of residue available for dry and moist hand wipes, respectively. By 120 hours post-treatment, transfer of carbaryl residue to hand presses corresponded to <0.001% of the theoretical amount of residue in all plots.

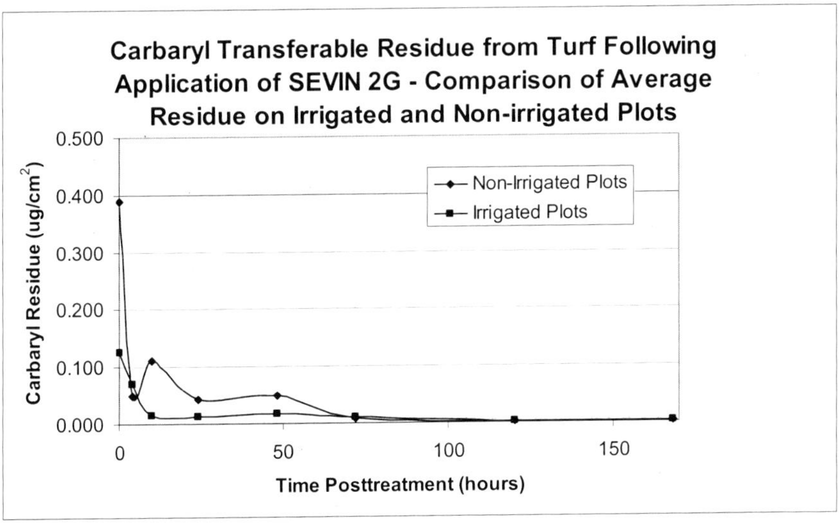

Figure 1. Comparison of average residues on cloth dosimeters collected from non-irrigated and irrigated turf plots following treatment with Sevin® 2G at a target rate of 9 lbs formulated product/1000 ft^2.

Table III. Carbaryl Residues from Hand Wipes from Hand Presses Collected from Turf Treated with Sevin® 2G

	Carbaryl Residue (μg/cm^2)			
	Non-Irrigated		*Irrigated*	
Days Post-Treatment	*Dry*	*Moist*	*Dry*	*Moist*
0	0.242	0.583	0.064	0.083
5	<0.020	<0.020	<0.020	<0.020

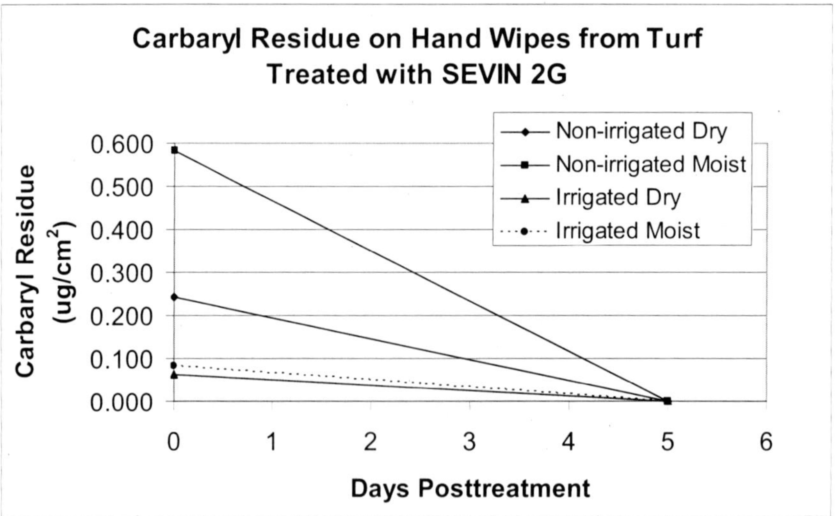

Figure 2. Comparison of average residues on hand wipes from moist and dry hand presses collected from non-irrigated and irrigated turf plots following treatment with Sevin® 2G at a target rate of 9 lbs formulated product/1000 ft^2.

Table IV. Carbaryl Residues from Hand Wipes from Hand Presses Collected from Turf Treated with Sevin® 2G Expressed as a Percentage of Residue Applied

	Percent Applied			
	Non-Irrigated		*Irrigated*	
Days Post-Treatment	*Dry*	*Moist*	*Dry*	*Moist*
0	0.275	0.662	0.065	0.095
5	<0.001	0.001	0.001	0.001

Conclusions

Following treatment with Sevin® 2G at a target rate of 9 lbs formulated product/1000 ft^2 (equivalent to 0.18 lb ai/1000 ft^2 or 8.8 kg ai/ha), cloth dosimeter samples were collected at various time points from three different trials using a Modified California Roller technique. Additionally, in one of the trials, moist and dry hand wipes were collected from hand press samples immediately following treatment, and 120-hours post-treatment.

The carbaryl residues measured on cloth dosimeters were lower in samples from the irrigated plots compared to corresponding samples collected from the non-irrigated plots. Measured carbaryl residues rapidly declined over the first 4 hours and then leveled off or rose slightly after 10 hours. At later time points, the transferable residues of carbaryl were very low.

As with the cloth dosimeters, initial carbaryl residues from both dry and moist hand wipes were considerably lower in hand press samples from the irrigated plot compared to similar hand press samples collected from the non-irrigated plot. For both non-irrigated and irrigated plots, hand wipes from moist hand presses had approximately twice the carbaryl residues as those from the corresponding dry hand presses. All hand wipe samples collected from hand presses 120 hours after treatment had residues <LOQ.

References

1. *Occupational and Residential Exposure Test Guidelines, Group B: Post-Application Exposure Monitoring Test Guidelines*; United States Environmental Protection Agency, Office of Pesticide Programs, 1996.
2. *40 CFR 160 Good Laboratory Practice Standards;* United States Environmental Protection Agency, Office of Pesticide Programs.
3. *40 CFR 792 Good Laboratory Practice* Standards; United States Environmental Protection Agency, Office of Prevention and Toxic Substances.
4. *ORETF-10.E.0.* Outdoor Residential Exposure Task Force, 1998.

5. *USEPA Project Summary EPA/600/SR-97/119*; Fortune, C. R., 1998.
6. *Assigning Values to Non-Detected/Non-Quantified Pesticide Residues in Human Health Food Exposure Assessments*; United States Environmental Protection Agency, Office of Pesticide Programs, 2000. USEPA Docket #OPP-00570A.

Chapter 5

Surface Drinking Water Assessment and Monitoring for Oxadiazon Herbicide on Golf Courses

Ujjana B. Nandihalli, Russell L. Jones, Richard Allen, Tharacad S. Ramanarayanan, and George J. Sabbagh

Bayer CropScience, 17745 South Metcalf, Stilwell, KS 66085

Modeling and drinking water monitoring studies were conducted to determine potential dietary exposure to oxadiazon as a result of its use on golf courses. As part of their Reregistration Eligibility Decision, the United States Environmental Protection Agency (USEPA) performed a worst case exposure estimation using the PRZM/EXAMS Index Reservoir model, assuming a maximum oxadiazon application rate (8.96 kg a.i/ha) and resulting in an acute potential exposure of 181 ppb and a long term mean of annual concentrations in drinking water of up to 56 ppb. A refined drinking water exposure assessment was conducted by the authors using the same modeling tools, but adopting more realistic assumptions reflecting that 90% of the product is used as a granular formulation, with typical use rates of 6.72 kg a.i./ha. In addition, a GIS evaluation of land use in Florida determined that golf courses represented a maximum of 6% of the surface area in watersheds. These and other refinements resulted in reductions in the estimated concentrations of over two orders of magnitude. To confirm the refined exposure assessments, a three-year surface water monitoring program was established in Florida and North Carolina to measure the potential for oxadiazon to reach surface drinking water sources in three community water systems, with the highest use of oxadiazon. Residues were detected in raw water in two of the three community water systems and in finished water in one. However the maximum observed concentrations from the monitoring program were more than three orders of magnitude

> lower than the acute drinking water concentration originally estimated and about 25 times lower than the refined assessment. These differences are associated with the assumptions concerning actual use in the watershed, spray drift and persistence of residues in the reservoir. The monitoring program also demonstrated that oxadiazon can be removed in drinking water treatment systems.

The Food Quality Protection Act (FQPA) mandated that risk contributions from the presence of pesticides in drinking water should be included in the overall human health risk assessment. The USEPA has been utilizing predictive models for estimating potential upper bound concentrations of pesticides in surface waters used as a source of drinking water. The linked PRZM/EXAMS model is commonly used by the agency as part of the screening process to assess the potential for drinking water-related exposures that may exceed the human health level of concern. Oxadiazon is an herbicide which is used on golf courses to control grassy weeds. This chapter describes (a) the drinking water exposure assessment for oxadiazon by USEPA, (b) the refined exposure assessment by the authors, and (c) surface water monitoring for oxadiazon residues.

Drinking Water Exposure Assessment by USEPA

As part of a Reregistration Eligibility Document (RED), the USEPA performed a Tier II PRZM/EXAMS simulation using the Florida turf scenario, the Index Reservoir and the Percent Crop Area (PCA) adjustment to predict concentrations of oxadiazon in surface water that serve as a source of drinking water (*1*). Input parameters are summarized in Table I. PCA adjusting factors from the EFED guidance document for the turf scenario were used (*2*). The PCA values were 0.04 for golf course greens and tees, 0.23 for fairways, and 0.67 for roughs. The predicted concentrations from the model were multiplied by the PCA factor to arrive at the final Estimated Drinking Water Concentrations (EDWC) for each segment of turf. An example calculation is shown below (EEC = Estimated Environmental Concentration):

$$\text{Cancer Chronic EEC} = (\text{Mean of annual value}) \times (\text{PCA tees \& greens})$$
$$= (52.56\ \mu g/L) \times (0.04) = 2.1\ \mu g/L$$

USEPA calculated EDWC values in surface water are as shown in Table II.

Table I. Simulation Paramaters Used by USEPA

Parameters and Units	*Value Used*
Molecular weight (g mole-1)	345.2
Vapor pressure (torr)	1.0 E-6
Water solubility (mg L-1)	1.0†
Hydrolysis half-life at pH 5 (days)	Stable
Hydrolysis half-life at pH 7 (days)	Stable
Hydrolysis half-life at pH 9 (days)	38
Aerobic soil metabolism half-life (days)	841*
Aerobic aquatic metabolism half-life (days)	1682**
Anaerobic aquatic metabolism half-life (days)	365
Direct aqueous photolysis (days)	2.75
Soil water partition coefficient, Koc (L kg-1)	2352
Pesticide application rate, first application (kg ai/ ha)	2.24
Pesticide application rate, second application (kg ai/ ha)	2.24
Pesticide application rate, third application (kg ai/ ha)	4.48
Date of first pesticide application	March 15
Interval between first and second pesticide application (days)	30
Interval between second and third pesticide application (days)	135
Spray efficicieny (percent)	99
Spray drift (percent)	6.40

† Measured water solubility was multiplied by 10 according to (*2*).

* Meaured value was multiplied by 3 according to (*2*).

** Used two times the input value for the soil aerobic metabolism half-life according to (*2*).

Table II. Estimated Drinking Water Concentrations for Oxadiazon Calculated by USEPA

Exposure	*Greens & Tees*	*Fairways*	*Roughs*	*Golf Course*
Acute (90th percentile)	7.7	44.3	128.7	180.6
Non-Cancer Chronic (90th percentile annual value)	2.8	15.9	46.2	64.9
Cancer Chronic (mean 36-year annual concentration)	2.1	18.6	35.2	56.0

Note: Units are µg/L (ppb)

Source: Reference *1*.

Refined Drinking Water Exposure Assessment

A refined drinking water exposure assessment was conducted by the authors using the same modeling tools as the USEPA, but adopting more realistic assumptions for product use. Oxadiazon is predominantly applied to golf course tees, greens, and fairways and therefore inclusion of roughs in the model would overestimate the predicted concentration. Over 90% of oxadiazon product is applied in granular formulation, and therefore spray drift should not be considered as a contributing factor. Also, a typical annual oxadiazon application rate is 6.72 kg. a.i/ha or less instead of the 8.96 lb. ai/ha that was applied in the USEPA assessment. Finally, the PCA assumptions used in the USEPA model are believed to be unrealistic, in the sense that they assume that 100% of the watershed area is covered with golf course.

In the refined assessment, PCA factors were computed for surface water intakes in Florida using the Geographic Information System (GIS). Florida has the highest concentration of golf courses in the country and GIS technique would provide the most likely maximum PCA factor for golf course. The PCA factors were calculated in two different ways:

- Land Cover: Area of recreational grasses class (includes golf courses) from National Land Cover Datasets (NLCD), divided by the watershed area.
- Golf Course Yardage: Areas of fairways in the watershed divided by the watershed area (area was calculated by mulitpling the yards of fairways in the watershed and an average width of 25 yards and then an additional 25 percent of the total was added to be conservative).

Golf courses in Florida and their yardage were identified from two sources, including the CNN/Sports Illustrated website. The maximum PCA was computed to be 5.77% (*3*).

The EEC results (i.e. EEC_{unadj}) predicted by PRZM/EXAMS for each exposure were multiplied by PCA factors as follows.

Golf course:	$[EEC_{unadj}] \times 0.0577$
Greens and tees:	$[EEC_{unadj}] \times 0.0577 \times 0.04$
Fairways:	$[EEC_{unadj}] \times 0.0577 \times 0.23$
Greens, tees, and fairways:	$[EEC_{unadj}] \times 0.0577 \times 0.27$

The EDWCs from USEPA assessments were compared with the refined assessment for water supplies in Florida associated with oxadiazon use on golf course (Table III). The values represent the sum of greens, tees, and fairways only; oxadiazon application to roughs is insignificant.

Table III. Comparison of EDWCs Between USEPA Assessment and Refined Assessment

Simulation	*Acute*	*Non-Cancer*	*Cancer Chronic*
USEPA simulation (8.96 kg ai/ha)	52.0	18.6	20.7
USEPA assessment with refined PCA	3.00	1.08	0.82
Refined assessment without drift (8.96 kg ai/ha)	2.62	0.91	0.61
Refined assessment without drift (6.72 kg ai/ha)	2.23	0.62	0.44

Note: Units are in µg/L (ppb).
Source: Reference 3

Surface Water Monitoring

To validate the predictions of the refined assessment, a surface water monitoring program with two sites in Florida and one site in North Carolina was instituted. The principal criteria used in the selection of sites were product sales by state, geographical distribution within the state and percent crop area within watersheds. Based on 2000-2001 oxadiazon sales by state, 27.8% of total sales in the U.S. occurred in Florida and North Carolina.

The Florida sites were chosen because Florida is the state with the highest sales of RONSTAR® herbicide. Two community water systems (CWS) were chosen in Florida, the West Palm Beach CWS on the east coast and the City of Bradenton CWS on the west coast. These two locations are representative of community water system watersheds with a high intensity of golf courses, and they are the only sites where the watersheds are used continually to supply water and the PCA of golf courses exceeded one percent with both estimation procedures.

The West Palm Beach CWS draws water from Clear Lake in Palm Beach County, which is supplied by a system consisting of a catchment area which captures rainfall and also stores water pumped from M-canal, which flows from Lake Okeechobee. A map showing intake location, golf course distribution and hydrology is presented in Figure 1.

The Bradenton CWS draws water from Lake Ward in Manatee County, which is a relatively small reservoir on the Bradenton River. Specific watershed data and golf course information are presented in Table III. A map showing intake location, golf course distribution, and hydrology is presented in Figure 2.

The Thomasville CWS in North Carolina was selected because it had the second highest PCA in the region. The City of Thomasville water treatment plant draws water from Tom-A-Lex Lake. A map showing intake location and golf course distribution is presented in Figure 3.

Specific watershed data and golf course information for all three sites are presented in Table IV.

Sampling and Residue Analysis

Raw surface water and finished drinking water samples were collected either weekly (Bradenton and Thomasville) or bi-monthly (West Palm Beach). Initially, raw water samples were analyzed and finished water samples were analyzed only when oxadiazon residues were detected in the corresponding raw water sample. The residue analysis was performed by LC/MS/MS for oxadiazon parent only. The method detection limit was 0.01 ppb (µg/L) and the limit of quantification was 0.03 ppb.

Table IV. Specific Watershed Data

Intake	*Treatment Plant*	*Land Cover PCA (%)*	*Yardage PCA (%)*	*Watershed Area (ha)*	*Number of Golf Courses*
Lake Ward at Bradenton, FL	City of Bradenton	5.54	1.24	14,200	5
Clear Lake at West Palm Beach, FL	City of West Palm Beach	4.37	1.03	52,300	21
Tom-A-Lex Lake at Thomasville, NC	City of Thomasville	Not Available	0.46	14,600	4

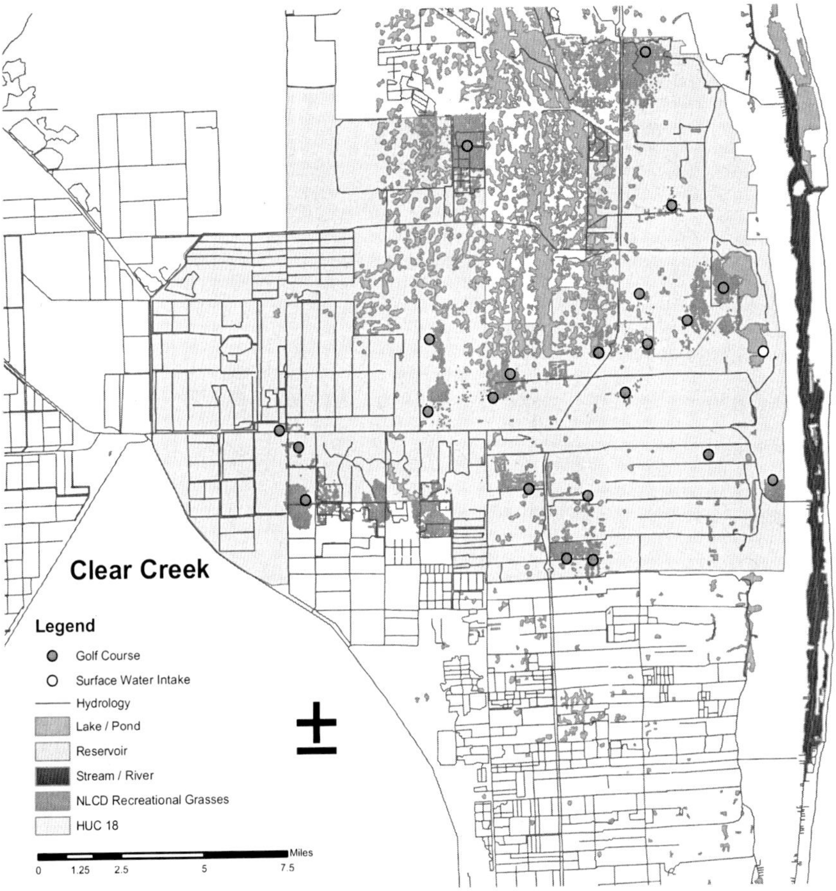

Figure 1. Watershed map of West Palm Beach Community Water System. (see page 1 of color insert)

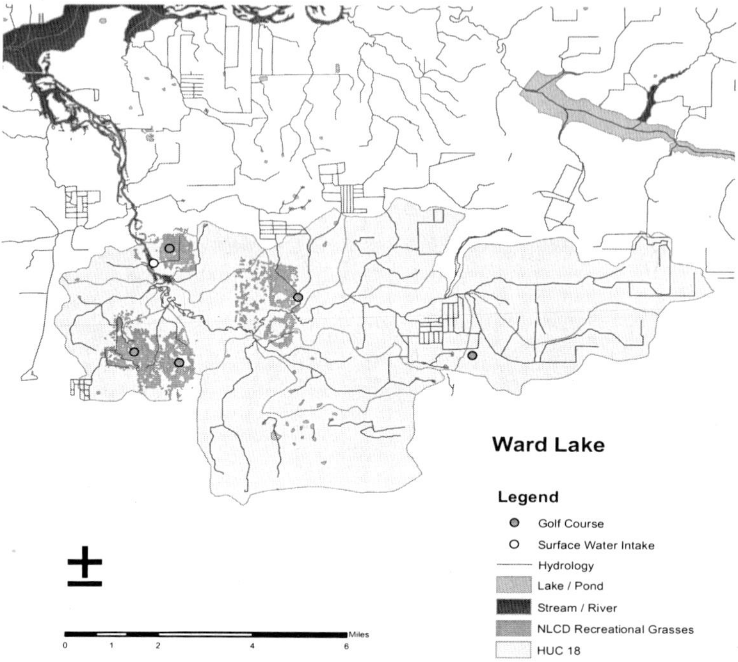

Figure 2. Watershed map of Bradenton Community Water System.
(see page 2 of color insert)

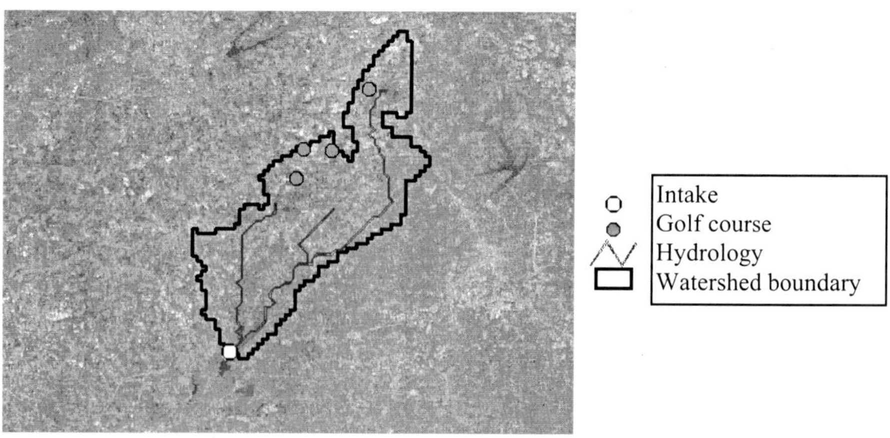

Figure 3. Watershed map of Thomasville Community Water System.
(see page 3 of color insert)

Results

The results of the three year monitoring program are summarized in Table V by year and by community water system. Residues of oxadiazon were not detected in raw water in the West Palm Beach community water system. Residues of oxadiazon were present in the raw water of the Bradenton community water system, but not in the finished. Residues of oxadiazon were present in both raw and finished water in the Thomasville community water system. The finding of residues in finished water in Thomasville but not Bradenton likely results from an activated carbon treatment step that is included at Bradenton but not at Thomasville. Figures 4 and 5 show the oxadiazon concentrations as a function of time at Bradenton and Thomasville.

Table V. Summary of Monitoring Results

		Peak Residue Concentration (ppb)		*Time-Weighted Average (ppb) in*
Site Name	*Year*	*Raw Water*	*Finished Water*	*Finished Water*
	1	0.059	0.005*	0.005*
Bradenton, FL	2	0.175	0.005*	0.005*
	3	0.086	0.005*	0.005*
	1	0.005*	Not analyzed**	Not applicable
West Palm Beach, FL	2	0.005*	Not analyzed**	Not applicable
	3	0.005	Not analyzed**	Not applicable
	1	0.170	0.127	0.025
Thomasville, NC	2	0.049	0.047	0.013
	3	0.051	0.055	0.015

* Residues were non-detectable (i.e. <MDL). One half the MDL (0.005 ppb) was used for reporting purposes. Similarly, one half the MDL was used in the TWA calculation.

** Finished water samples were not analyzed because there were no detectable residues in the corresponding raw water samples.

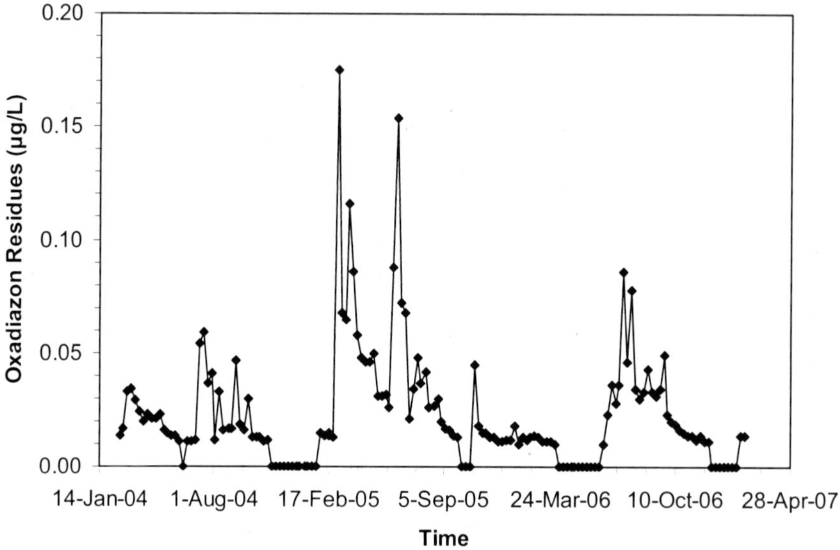

Figure 4. Oxadiazon Residues in raw water samples from Bradenton (there were no residues in finished water).

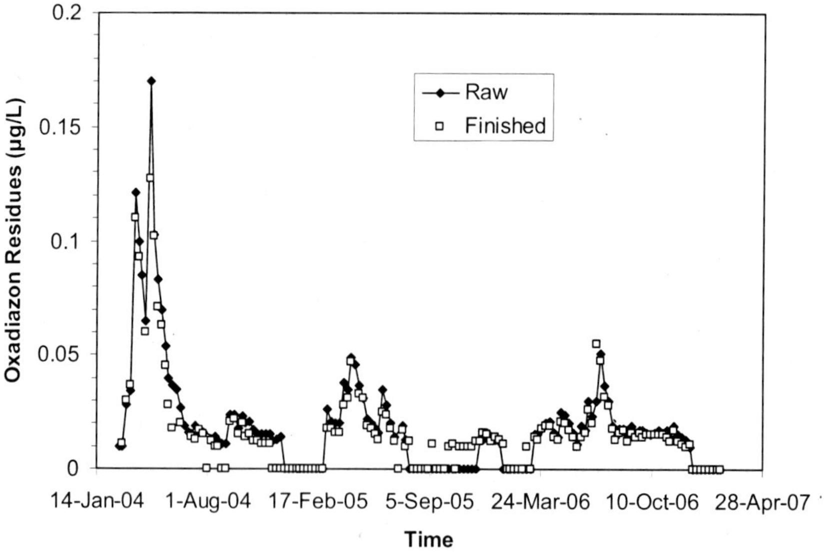

Figure 5. Oxadiazon residues in raw and finished water at Thomasville.

Table VI compares the exposure assessments and the monitoring study. The worst case exposure estimation by USEPA using the PRZM/EXAMS Index Reservoir model assumed a maximum application rate of 8.96 kg a.i./ha applied through ground spray, resulting in an acute exposure of 181 ppb and a long term mean of annual concentrations in drinking water of up to 56 ppb. The refined drinking water exposure assessment, using the same modeling tools but adopting more realistic assumptions as to product formulation and PCA factors, resulted in reductions in the estimated concentrations of over two orders of magnitude. In the surface water monitoring study, residues were detected in raw water in two of the three monitored community water systems, and in finished water in one, with the highest peak and annual mean concentrations being lower by more than an order of magnitude than the refined assessment.

Table VI. Comparison Between Simulated and Measured Concentrations

Case	*Peak Residue*	*Annual Mean Residue*
USEPA Assessment	52.0	18.7
Refined Assessment	2.23	0.62
Monitoring Study	0.127	0.025

Note: Units are in μg/L (ppb)

Discussion

The results show how conservative assumptions can propagate through a risk assessment and result in overly conservative estimates of exposure. One of the most conservative assumptions in the USEPA assessment is the estimation of use intensity in a watershed. Work done by the United States Geological Survey (USGS) (*4*) has shown that use intensity (the amount of active ingredient applied divided by the watershed area) is by far the most important variable affecting residues in surface water which may arise following pesticide applications to agricultural fields. Use intensity is usually calculated by multiplying the treated area by the average annual application rate, or there may be crop protection sales figures available for a specific watershed. The assumptions in the USEPA risk assessment overestimate risk by assuming that all, or almost all, of the watershed is composed of golf courses and that all of the target area is treated at the maximum rate. For this compound, one of the most conservative assumptions was that regarding the percentage of the watershed area in turf on golf courses. The USEPA value was a factor of 17 higher than that determined by a GIS analysis. The assumption that all of the area was treated with oxadiazon using the maximum number of applications and the maximum label rate also overestimated the amount used in the watersheds. There are a number of other conservative assumptions in the USEPA assessment, including conservative estimates of oxadiazon properties, a simplified watershed hydrology, the assumption that all golf courses immediately discharge runoff water after a rain (many golf courses store water in ponds for irrigation), the assumption that all oxadiazon was applied as a liquid when over 90 percent is applied as a granular formulation, and no accounting for the removal of residues during water treatment. Therefore, the almost three order of magnitude difference between the USEPA assessment and the actual residues measured is not surprising. Because of more realistic assumptions, the refined assessment did a better job of predicting concentrations in actual watersheds. However, the refined assumptions regarding use, watershed hydrology, recirculation of water and removal during water treatment were still conservative, so the predicted exposure concentrations were still a factor of 25 higher than actually observed.

The overly conservative predictions of the USEPA procedures are not limited to products used on golf courses. Drinking water monitoring studies conducted by Bayer CropScience (*5*) with four compounds, all with agricultural uses but also including one with home and garden uses, and targeting watersheds with the highest use intensities, showed similar differences between predicted and observed concentrations. A similar conclusion was presented in a paper by industry scientists (*6*) comparing predictions with the results observed for a large number of compounds in an USEPA/USGS reservoir monitoring program.

Conclusions

- The USEPA estimates of oxadiazon concentrations in drinking water obtained from a surface water source were overly conservative, primarily due to assumptions about the maximum PCA, but also with respect to assumptions regarding spray drift and application rate.

- The maximum observed annual average concentration of oxadiazon from the monitoring program was almost three orders of magnitude lower than originally estimated by USEPA, and about 25 times lower than the refined exposure assessment.

- Oxadiazon is removed in drinking water treatment systems, with complete removal occurring when carbon treatment is included.

References

1. USEPA Memorandum, April 15, 2002, PC Code: 109001, DP Code: D281176.
2. USEPA EFED Guidance Document, 2001. Version II, December 4, 2001.
3. Andrish, S.D.; Kay, S.A.; Snyder, N.J. 2003. Waterborne Environmental Inc., WEI 746.29, MRID #45920102.
4. Crawford, C. G., S. J. Larson, R. J. Gilliom. 2005. United States Geological Service, Indianapolis, Indiana
5. Jones, R. L. *J. Agric. Food Chem.* **2005**, 53, 8835-8839.
6. Jackson, S.; Hendley, P.; Jones, R., Poletika, N.; Russell, R. *J. Agric. Food Chem.* **2005**, 53, 8840-8847.

Chapter 6

Comparison of Regulatory Method Estimated Drinking Water Exposure Concentrations with Monitoring Results from Surface Water Drinking Supplies

Scott H. Jackson[1], John J. Jones[1], John Hanzas[2]

[1]BASF Corporation, 26 Davis Drive, Research Triangle Park, NC 27709
[2]Stone Environmental Inc., 535 Stone Cutters Way, Montpelier, VT 05602

Plant-protection compounds are often necessary to maintain the value and aesthetics of high value turf grass. However, these products can enter aquatic systems either by direct or indirect means. In order to better understand the possible frequency and magnitude of exposure to drinking water systems, golf courses were identified within watersheds that have historically been treated with vinclozolin fungicide. Golf courses selected in this work were chosen based on their geographic distribution, representation of the product use area, and course position in a watershed relative to a surface water intake(s). Additionally, wells were identified adjacent to courses treated with vinclozolin so that the extent of possible exposure to ground water could also be determined. Analytical results from the study were then compared via several exposure estimation methods to evaluate the ability of the modeling methodologies to predict concentrations found in drinking water based on monitoring data. Results from this examination indicate that exposure estimates based on models can overpredict concentrations found in water by several orders of magnitude.

Introduction

Plant protection products are applied to high value turf grass for a variety of reasons, including controlling weeds, insects and pathogens. Installing and maintaining high value turf is an expensive activity, and therefore managers work to protect the investment using chemical products. One of the goals of site managers is to use products responsibly, so that they do move into non-target areas. However, in order to evaluate the potential for compounds to enter ground and surface water, the regulatory community may require registrants to conduct environmental monitoring. Regulators also typically use exposure models to help determine the possible impact of off-target movement of crop protection chemicals into water resources. Therefore, unless the models relied upon are able to properly characterize the environment they are meant to describe, improper conclusions may be drawn regarding the fate of those products in the environment. To date, monitoring chemical exposure from golf courses has principally focused on edge of green or fairway losses, and not on which residues may possibly reach a stream system transporting water to a drinking water supply (*1-6*). Vinclozolin is an example of one of the compounds used by turf managers to maintain high value turf. Vinclozolin is a non-systemic fungicide used for the control of Botrytis spp., Sclerotinia spp., Monilia fruiticola and Gloeosporium spp.

The authors had previously reported to regulatory authorities that several exposure estimate methods indicated that use of vinclozolin on turf would pose little risk to drink water sources. However, interest was expressed in the collection and evaluation of monitoring data. Therefore, the work reported in this study was initiated as the result of a regulatory request. The focus of this effort was to generate vinclozolin monitoring data from Ohio and Pennsylvania locations to characterize the potential of exposure of drinking water supplies. The states of Ohio and Pennsylvania were chosen for this monitoring study based on vinclozolin sales data. Sales data obtained from Dmrkynetec (Doanes Marketing Research) indicated that Ohio and Pennsylvania were two of the highest use states for vinclozolin-containing products. A monitoring project was conducted to provide an added level of certainty that the chemical was not impairing water resources. The focus of this manuscript is to evaluate several modeling methodologies and compare them to actual monitoring data generated to characterize potential vinclozoling exposure to drinking water supplies.

Method and Materials

When a risk assessment is conducted, the starting point is typically an exposure model, since monitoring data are non existent, and there are typically not other data and methods available for refining risk. In order to understand the process whereby monitoring studies are often requested, it is necessary to first look at standard scenario modeling. When questions exist about potential crop protection chemical exposure to drinking water systems, models are used to estimate possible concentrations in the water systems. Jackson et al. (*7*) suggested that model predictions using the United States Environmental

Protection Agency (USEPA) standard scenarios could be as high as six orders of magnitude greater than actual water concentrations. Figure 1 is an illustration of a conceptual model for refining exposure estimates. USEPA has chosen to use a 1 in 10 year exposure event based on modeling predictions to characterize assessments (their regulatory protection goal). The question is, how much monitoring data needs to be collected to achieve the same protection goal standard as used with exposure modeling, so that they are directly comparable? A panel of experts at a workshop sponsored by the International Life Sciences Institute (ILSI) on water monitoring study design concluded that by having an adequate number of monitoring locations with adequate geographic representation, it was possible to achieve the same protection standard as a 1 in 10 year predicted exposure within a one year time frame (*8*). Therefore, a properly designed monitoring study could in one year provide data that was as equally protective as a 1 in 10 year modeled value.

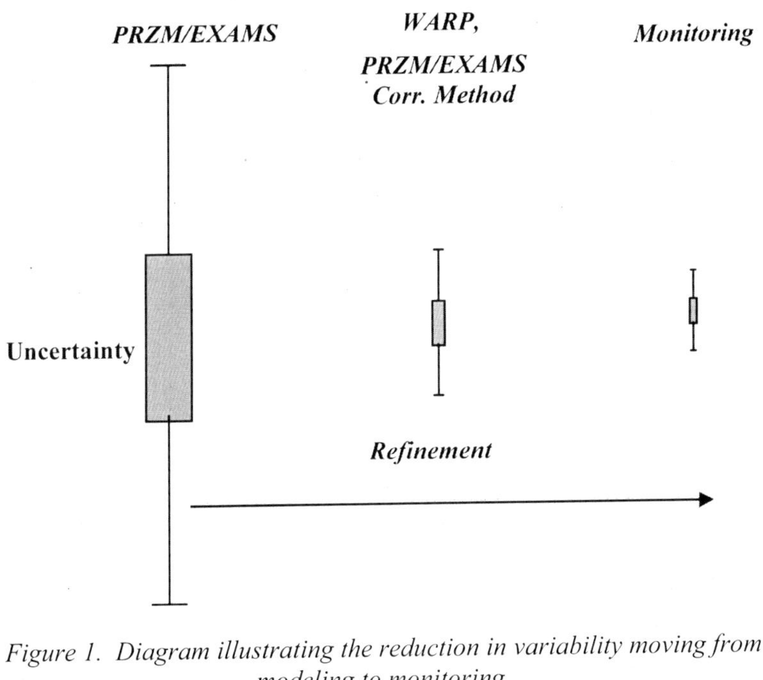

Figure 1. Diagram illustrating the reduction in variability moving from modeling to monitoring.
(see page 3 of color insert)

Using Calculation Methods to Estimate Exposure

Several calculation and modeling methods are presented in this section:

1. A Geographic Information System (GIS) watershed dilution based method using BASF sales data
2. A GIS watershed dilution based method using maximum application rate and seasonal load data
3. USEPA's standard Pesticide Root Zone Model/Exposure Analysis

Modeling System (PRZM/EXAMS) turfgrass scenario

4. USEPA's standard PRZM/EXAMS turfgrass scenario modified using the method of Jackson et al. (*7*)
5. The Watershed Regressions for Pesticides (WARP) model of Crawford et al (*9*). The sales date used for the BASF analysis were obtained from Dmrkynetec (*10*)

A brief description of each exposure estimation methodology as used in the author's analysis is presented below:

1. The GIS watershed dilution-based exposure estimation using BASF sales data was developed by first determining the total vinclozolin sales in the state of interest. The next step was to determine the number of golf course greens in the state, and then take the sales total and divide it by the total number of (state) greens to determine an application rate per hole in lbs/A. The next step was to determine the number of greens in any given watershed in the state to determine a total application rate per watershed (rate/hole x number of holes). We next determined the total June/July rainfall total in the watershed (~8") to determine a total water volume for the peak application period. The water concentration estimate was then determined by taking the total amount of active ingredient estimated for each watershed and dividing that mass by the total estimated water volume (e.g. ug/L).

2. The GIS watershed dilution-based exposure estimation method used by USEPA was similar to the method used by BASF, however rather than using actual sales data, the total amount applied in any watershed was based on the maximum allowable amount that could be applied, based on the product's label. Therefore, the first step was to determine the total amount of vinclozolin that could be applied per area, using 180 acres as the standard golf course area (1440 lbs/course). The next step was to determine the number of golf courses in a given watershed in the state to determine a total application rate per watershed (rate/course x number of courses). We next determined the total June/July rainfall total in the watershed (~8") to determine a total water volume for the peak application period. The water concentration estimate was then determined by taking the total amount of active ingredient estimated for each watershed and dividing that mass by the total estimated water volume (e.g. ug/L).

3. The next approach used to estimate potential exposure in the watershed was USEPA's standard PRZM/EXAMS turf grass scenario (*11-14*). The method is described in detail in the references cited.

4. The next approach used to estimate exposure in the watersheds was based on taking the USEPA's standard PRZM/EXAMS turf grass scenario estimate, and then applying a modifying factor to it based on the work of Jackson et al (*7*). The modifying factor is presented as follows:

$$y = 2.156 + 1.03584 * x$$

where:

y = log of model over prediction
x = log of total active applied

Taking 10y , or 1/ ≈ over prediction factor (OPF), the reprocical of OPF x P/E = Modified Value

Figure 14 presents the relationship between model over prediction and the correction factor used in the method.

5. The final exposure estimation method used was the WARP regression as developed by Crawford et al. (12). The WARP regression has been developed into several different versions and therefore we present the form of the equation we used, for clarity.

log10(concentration) = [(use intensity)1/4, log10(R-factor), K-factor, (watershed area)1/2, Dunne overland flow, May precipitation departure]+log10((SWMI/SWMIatrazine)1.125)+log10 ((vapor pressure atrazine/ vapor pressure)0.075)

where:

Use intensity	is the annual agricultural pesticide use in the watershed (kg) / watershed area (km^2) for the pesticide of interest
R-factor	is the rainfall erosivity factor from the Universal Soil Loss Equation
K-factor	is the soil erodibility factor from Universal Soil Loss Equation
Watershed area	is the area of the drainage basin (km^2)
Dunne overland flow	is the percentage of total stream flow derived from surface runoff caused by precipitation on saturated soil
May precipitation departure	is the departure from the 1961-90 average precipitation that occurred during the month of May during the one year of sampling (mm)
$SWMI_{vin}/SWMI_{atz}$	is the ratio of the surface water mobility index of the pesticide of interest to the sur-face water mobility index of atrazine
vapor $pressure_{atz}$/vapor $pressure_{vin}$	is the ratio of the vapor pressure for atrazine to the vapor pressure of the pesticide of interest.

Analytical Method

BASF Technical Procedure D0406, "Method for Determination of Vinclozolin (BAS 352 F) and Its Metabolites (BF 352-22, BF 352-23, BF 352-31, and BF 352-41) Residues in Water Using LC-MS/MS", was used for sample analysis. The method LOQ is 0.05 μg/L for BF 352-31 and 0.10 μg/L for all other analytes.

Analytical Method Summary

The analytical method is a common moiety method where BAS 352 F, BF 352-22, BF 352-23, and BF 352-41 are converted through a base hydrolysis step into BF 352-31 (3,5-dichloroaniline, DCA). An aliquot is removed from the water sample, 17N NaOH is added, and the sample is heated for two hours in a sealed culture tube. After hydrolysis, the samples are cooled and acidified with formic acid. The samples are then desalted using an Oasis® SPE column, eluted from the column with 1% formic acid in methanol and diluted to an appropriate volume with 1% formic acid in water for analysis via LC-MS/MS. Analyzing for the metabolite BF 352-31 alone may be accomplished by direct injection of an aliquot of the sample on the LC-MS/MS. This method was developed at BASF, Research Triangle Park, N.C., USA.

Procedural Recoveries

Each sample set contained at least one control (from residential wells in Southeast Wake County, NC) fortified with each analyte. The fortification levels were 0.50 or 0.05 ppb for BF 352-31 and 1.0 or 0.10 ppm for all other analytes. In addition, at least one unfortified control water sample was included with each analytical set. The summary of average procedural recoveries of vinclozolin and its metabolites is provided in Table I. The analyte structures can be found in Figure 3.

Table I. Procedural Recoveries of BAS 352 F and Its Metabolites from Fortified Well Water

Analyte	*Fortification Level (ppb)*	*Average Recovery (%) ± Standard Deviation*
BAS 352 F	0.1	95.4% ± 26.5% (n=24)
BF 352-22	0.1	102.5% ± 34.2% (n=24)
BF 352-23	0.1	106.0% ±26.6% (n=23)
BF 352-41	0.1	105.1% ± 27.3% (n=23)
BF 352-31	0.05	88.7% ± 23.1% (n=25)

Drinking Water Monitoring Study Design

Ten community surface water system sites used to supply drinking water to the public were identified in Ohio and Pennsylvania. These water systems were also determined to have the greatest contamination potential based on labeled uses of vinclozolin as applied to golf course turf. The criteria used to select the community water systems (CWS's) specified that there were golf courses draining into the watershed supplying water to the CWS; that the golf courses draining into the watershed supplying water to the CWS used vinclozolin; and that the highest ratio of vinclozolin treated golf course area to non-golf course area be determined and included. The methods used to select CWS's in each state are summarized in the following sections.

Surface Drinking Water System Watershed Delineation and Evaluation

In Ohio, a list of community water systems was obtained from the Ohio Environmental Protection Agency. This CWS database identified 136 unique systems which used surface water as the source for their drinking water supply. The database included system information such as state ID number, address, phone number, and source water and location of the system intake. A Geographic Information System analysis (GIS) was used to facilitate finding the location of system intakes. The GIS analysis was also used in conjunction with a hydrography dataset to visually delineate system watershed boundaries and size. Golf course locations (provided by *Golf Magazine*) were then overlaid using GIS, so that the number of golf courses within each watershed could be determined. The ratio of golf course to non-golf course area for each watershed was calculated to produce a vinclozolin relative vulnerability ranking. If a watershed's ratio of 0.01 or greater was determined, then the CWS was included in the golf course personnel interview process. If a watershed ratio of less than 0.01 was determined, then the system was no longer considered for selection in the study. In Pennsylvania, watershed delineation and evaluation were conducted using methodology similar to Ohio, except that the Pennsylvania Department of Environmental Protection would only provide the name of the CWS source water and not the actual intake locations. Therefore, all CWS intake locations were determined using GIS to find the centroid of the CWS zip code. The next step in the determination was to find the nearest point in the hydrography to the source water.

Golf Course Personnel Interviews

Personnel at each golf course in the watersheds with the highest area ratio of courses were contacted between January and April 2004 to determine if they included vinclozolin in their turf management practices. During the interviews, questions were asked such as what the fungicide use at each course was for the previous two years, and what application plans for the upcoming year were. Additionally, course personnel were asked for information on application timing, the size (area) of their course, and where product was typically applied (greens, tee boxes and/or fairways) on the course. The exact geographic course location within each watershed of interest was also confirmed during the interviews. If the course intended to use vinclozolin during the next season, then the course was included in a re-ranking process. Application timing information was used to plan the sample collection initiation and frequency. Once golf courses confirmed vinclozolin use, community water supplies could be selected within the watersheds that had the highest amount of compound applied.

The next step in the process was to conduct interviews with water plant superintendents or operators to determine their willingness to participate in a monitoring program. As part of the interview, plant superintendents or operators were asked to confirm their contact information, source water name, and intake location. The confirmation of intake location was crucial for ensuring that the correct watershed was being evaluated. Additional interview information included the population size of the area served, details regarding water treatment methodology, and whether the plant solely used surface water for supplying the community served (e.g., that water was not blended with ground water sources). Systems were removed from consideration if they did not actually use surface water, if they did not solely use surface water, or if they were not interested in participating in the sampling program. Watershed/CWS systems were then re-ranked based on survey information.

Final Watershed Delineation

Water plants in watersheds determined to use maximum amounts of vinclozolin, that solely used surface water and were willing to participate in the sampling program received a refined watershed delineation using GIS. The watershed delineation process followed the following steps: all Hydrologic Unit Code (HUC) boundaries upstream of the system intake point were combined until all source water was accounted for. In Ohio, HUC12s were used for this process while in Pennsylvania HUC14s were used. The watershed boundaries from the intake point to the HUC boundaries were delineated to complete the watershed. The delineation was performed using a 1:24,000 scale digital topographic map for reference.

Ground Water Site Selection Methodology

Ten (total) wells that supplied potable water and were located adjacent to golf courses were identified in eastern Ohio and western Pennsylvania. The sites were also identified as potentially vulnerable to contamination from vinclozolin use at golf courses. The strategy used for selection of the wells was to:

1. identify areas where the ground water aquifer was vulnerable to contamination from surface applied chemicals
2. contact golf courses in these areas to confirm vinclozolin use
3. locate vulnerable wells down-gradient from the golf courses where vinclozolin had been used in these areas.

Further detail regarding the method used to select wells based on vinclozolin use and ground water vulnerability is provided in the following sections.

DRASTIC Scores

DRASTIC is an acronym for Depth to water, net Recharge, Aquifer media, Soil media, Topography, Impact of the vadose zone media, and hydraulic Conductivity of the aquifer. DRASTIC factors were weighted and combined to calculate a pollution potential rating, which was the DRASTIC score. The DRASTIC method identifies an area's potential ground water vulnerability to surface chemical exposure.

Ohio

A statewide glacial geology map obtained as an ArcView shapefile was used as an initial screening tool to identify counties in eastern Ohio that contained a large expanse of sand and gravel media-composed aquifers. Aquifers composed of sand and gravel media are considered the most vulnerable to pesticide contamination among major aquifer types found in the state. The areas that were determined to be most vulnerable were Summit, Stark, Portage, Medina and Wayne counties. The Ohio Department of Natural Resources (ODNR) was in the process of completing ground water pollution potential mapping for every county in the state using the DRASTIC methodology. In the Ohio well selection process, a variation of the general DRASTIC system, referred to as Pesticide DRASTIC, was used. Pesticide DRASTIC was developed by ODNR to more specifically account for factors that might help determine potential exposure from pesticides. Pesticide DRASTIC maps were obtained from ODNR of Summit, Stark, Portage, Medina, and Wayne counties as ArcView shapefiles.

Pennsylvania

The Pennsylvania Department of Conservation and Natural Resources (DCNR) had also created a statewide DRASTIC map that was used in the well search. A variation of the general DRASTIC system called Agricultural DRASTIC was developed by DCNR to more specifically account for factors that determine vulnerability to pesticide leaching in Pennsylvania.

Evaluation Process

For both the Ohio and Pennsylvania site selection processes, the county DRASTIC GIS shapefiles were brought into ArcGIS (Version 8), along with a nationwide data layer of golf course locations obtained from *Golf Magazine*. The relative vulnerability of the ground water in the area of each golf course was determined by overlaying the golf course locations on the DRASTIC data layers, and noting the areas of intersection. Upon reviewing the combined maps for the five Ohio counties, it was apparent that Summit and Stark counties had the most extensive areas of highly vulnerable ground water. Well selection was therefore focused on these two counties. Golf courses in Summit and Stark Counties were included on a list of candidate courses if the pesticide DRASTIC score in the area equaled or exceeded 160. According to the personnel of ODNR, a pesticide DRASTIC score of 160 or higher indicates moderately to highly vulnerable ground water conditions. Slightly more than half of the golf courses in Summit and Stark counties were eliminated based on the DRASTIC score thresholds because they were deemed not to be vulnerable enough. The remaining courses in Summit and Stark County were contacted to determine the possibility of participation in the vinclozolin well monitoring portion of the study.

The areas in western Pennsylvania with the largest extent of vulnerable ground water were in Mercer and Crawford counties. Therefore, golf courses in Mercer and Crawford Counties were included on a list of candidate courses if the DRASTIC score in the area equaled or exceeded 130. Although a score of 130 is a lower threshold than was used in Ohio, the scores are not directly comparable since the DRASTIC scores are calculated differently by the two states. Golf courses in Crawford and Mercer Counties were contacted about participation in the well monitoring portion of the vinclozolin study. Since there was a small number of candidate golf courses identified in Crawford and Mercer counties, the well search was expanded to all of Pennsylvania west of Pittsburgh, including Erie, Lawrence, Allegheny, Butler, Greene, Washington and Westmoreland counties. In Erie County, three golf courses were identified with DRASTIC scores exceeding 130, but all were found to be located in areas served by city water. The 21 remaining golf courses in Erie County had DRASTIC scores under 130. Twenty-nine golf courses were then identified in Lawrence, Allegheny, Butler, Greene, Washington, and Westmoreland Counties with DRASTIC scores of at least 124. Final well selection came from Crawford and Lawrence counties.

Golf Course Interviews

Golf courses with high pesticide DRASTIC scores were contacted between March and April 2004 to determine vinclozolin use and well locations. If the golf course superintendent or other grounds personnel indicated that vinclozolin products were not used on the course, the interview was ended and the course was excluded from further consideration. Golf courses were considered to merit further investigation if vinclozolin was applied in 2002 and/or 2003, and applications were planned for 2004. Golf course personnel were also asked about potable well locations on the course property and around the vicinity of the course. In many cases, the golf course clubhouse and other facilities were served by a municipal water system.

The seven golf courses identified as the best candidates in Pennsylvania and Ohio were visited in April 2004. Each course area was searched for wells that were in use and that were estimated to be hydrologically down-gradient of the golf courses. USGS 1:24,000 topographic maps were reviewed to estimate ground water flow patterns. Homeowners were interviewed regarding their willingness to participate in a monitoring program. Information regarding their well and water systems, including description of any treatment system present, was also collected. No more than two wells were selected at any golf course, to ensure spatial distribution of the study wells. Using information provided by the homeowner (e.g. the original homeowner's name), the state well databases were searched for well log information. Based on the state well database searches, well logs were found and evaluated for each of the selected sites. The wells determined to have the most shallow ground water and that were closest to the golf courses were selected. A total of seven wells in Ohio and three wells in Pennsylvania were selected for the monitoring program. More wells were selected in Ohio than Pennsylvania based on the vulnerability and use selection criterion. The wells selected for monitoring were considered among the most vulnerable wells in their respective areas.

After final site selection was completed, surface water samples were collected bi-weekly for the first six months following sampling initiation, with samples being collected monthly thereafter. Both raw and finished water were collected at each sample collection. Groundwater samples were collected from the wells quarterly, at a tap. Samples were unfinished or raw.

Rainfall During the Study Period

For each watershed, a representative weather station was chosen from the National Climatic Data Center (NCDC) cooperative weather station network. Monthly precipitation data were obtained from the designated station from April 2004 through March 2006. Rainfall was well above normal at 8 of the 10 sites during the study conduct. Rainfall was about 10 percent lower than average at two of the sites.

A summary of the community surface water supplies of the ten watersheds can be found in Table II. GIS coverage displaying the monitored watershed feeding the community water supplies can be found in Figure 2. GIS coverages of each watershed and associated community water supply are presented in Figures 4 to 13.

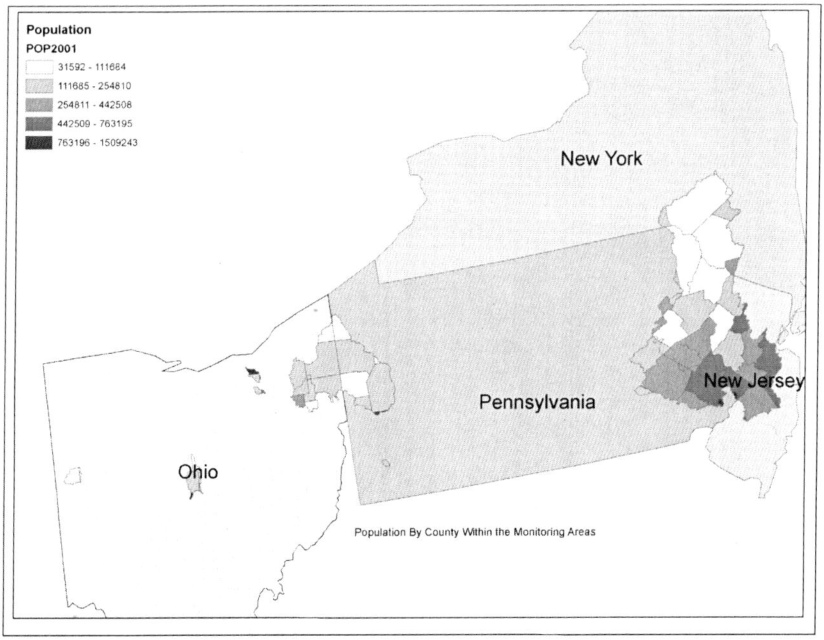

Figure 2. General location of the study watersheds within each state. Population within each watershed is symbolized.
(see page 4 of color insert)

Table II. Summary of the Study Watersheds With the Area Encompassed by Each in Square Miles

System ID	*State*	*Area of Watershed (sq. miles)*
OH-BA-VN	Ohio	28.58
OH-BE-VN	Ohio	64.14
OH-CE-VN	Ohio	112.32
OH-CH-VN	Ohio	195.72
OH-NF-VN	Ohio	305.03
PA-EL-VN	Pennsylvania	20.38
PA-NO-VN	Pennsylvania	1,736.99
PA-PBA-VN	Pennsylvania	8,903.13
PA-PBE-VN	Pennsylvania	1,874.93
PA-BF-VN	Pennsylvania	3,112.78

Vinclozolin

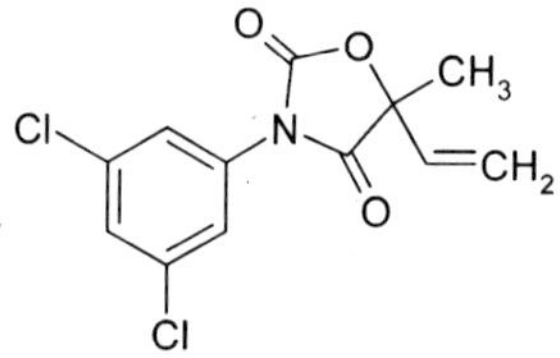

Degradates ~ 0.05 ug/L (LOQ) "LC-MS/MS"

(Met B)

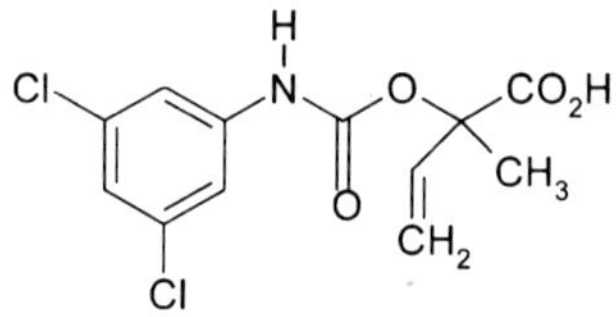

(Met E)

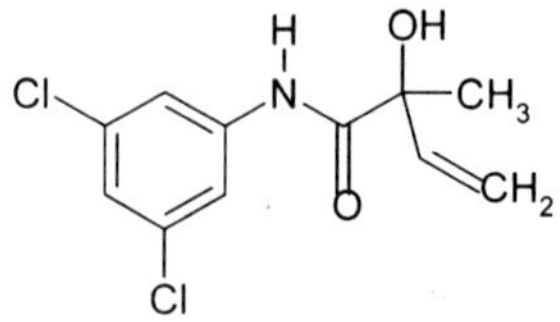

(Met D; DCA)

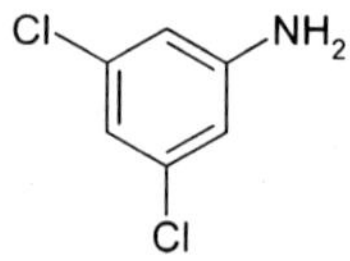

Figure 3. Structures of the analytes included in the analytical method.

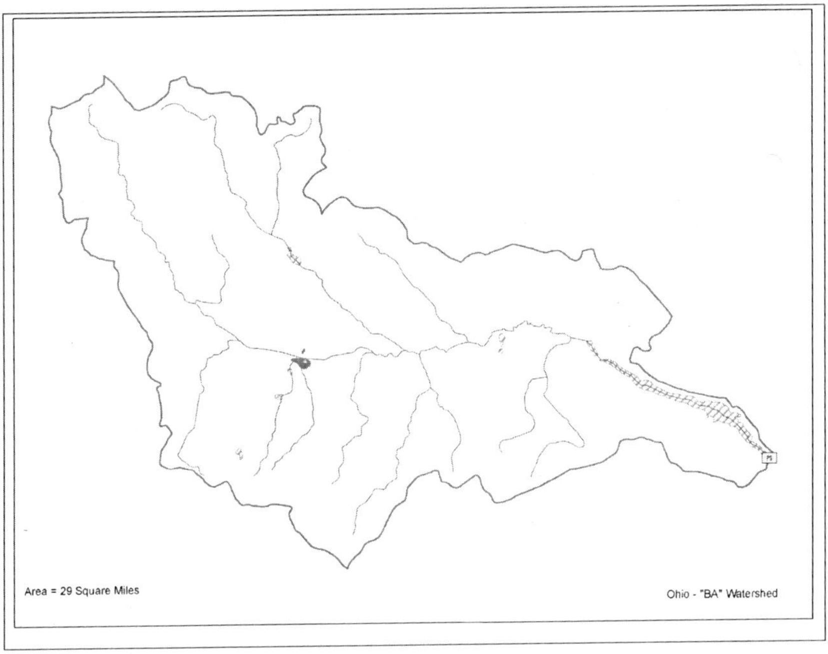

Figure 4. GIS presentation of the Ohio BA-VN watershed.
(see page 4 of color insert)

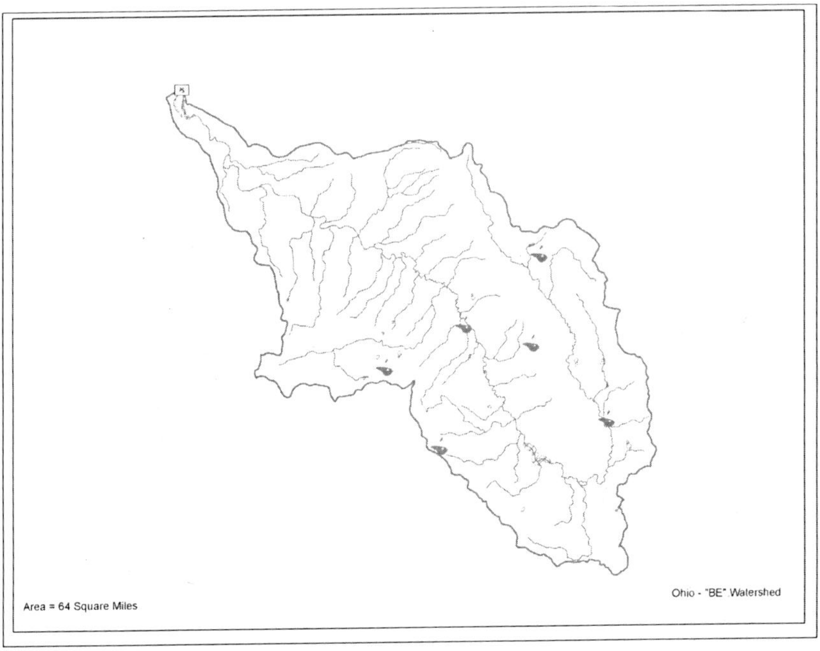

Figure 5. GIS presentation of the Ohio BE-VN watershed.
(see page 5 of color insert)

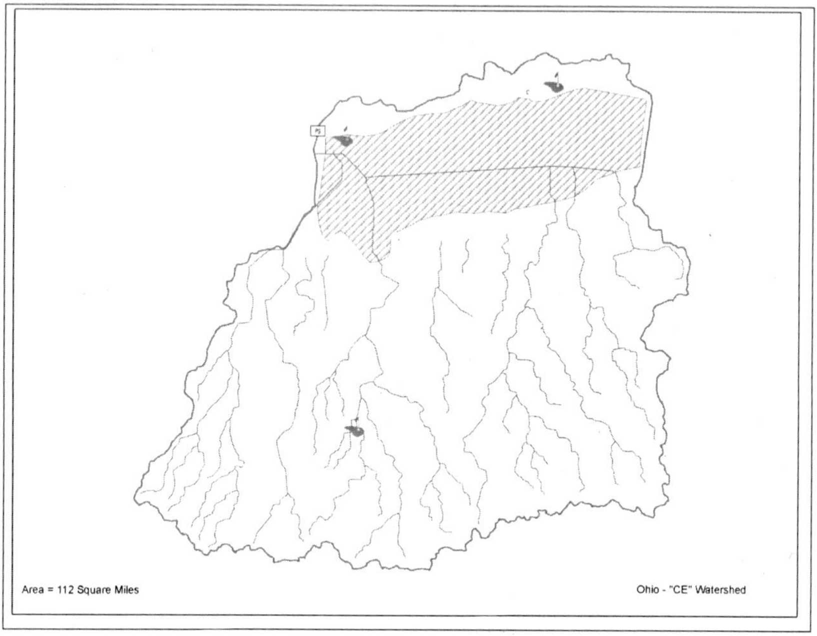

Figure 6. GIS presentation of the Ohio CE-VN watershed.
(see page 5 of color insert)

Figure 7. GIS presentation of the Ohio CH-VN watershed.
(see page 6 of color insert)

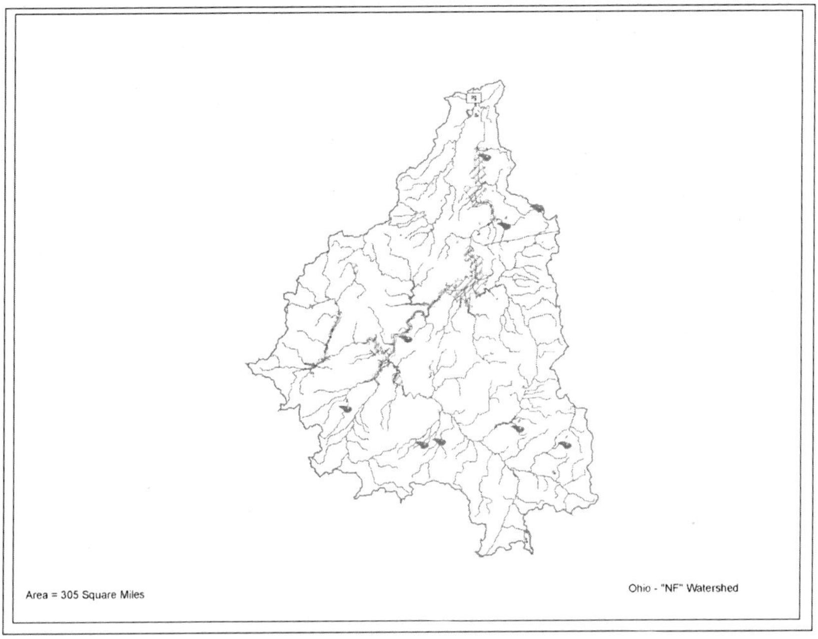

Figure 8. GIS presentation of the Ohio NF-VN watershed.
(see page 6 of color insert)

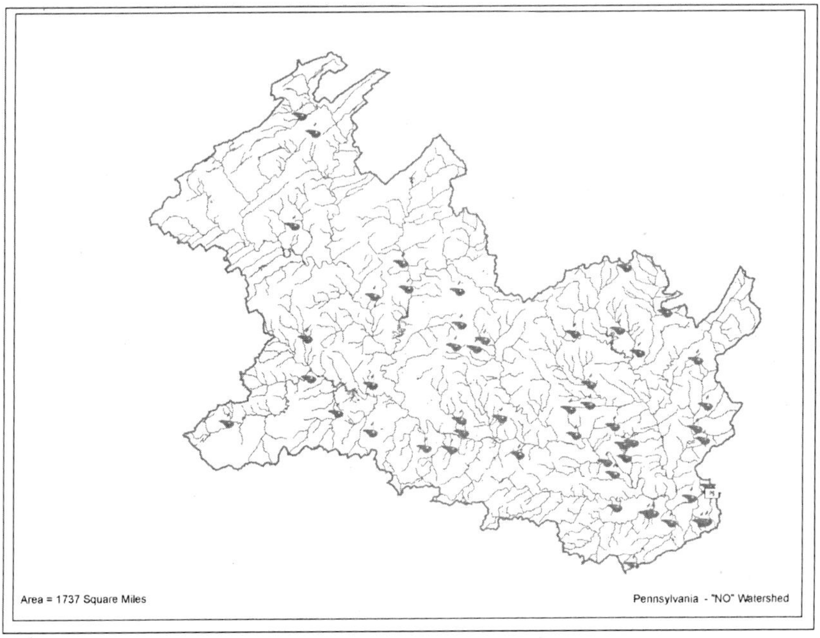

Figure 9. GIS presentation of the Pennsylvania EL-VN watershed.
(see page 7 of color insert)

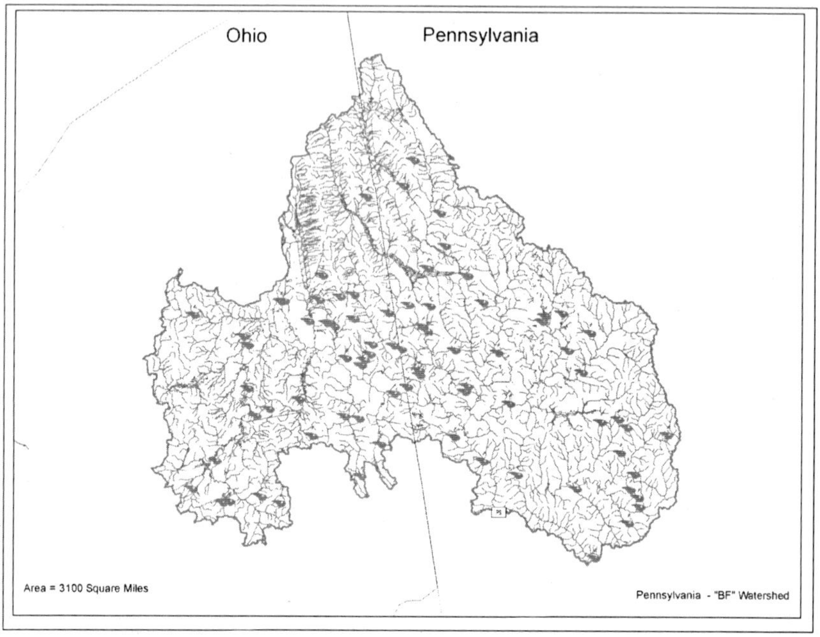

Figure 10. GIS presentation of the Pennsylvania NO-VN watershed.
(see page 7 of color insert)

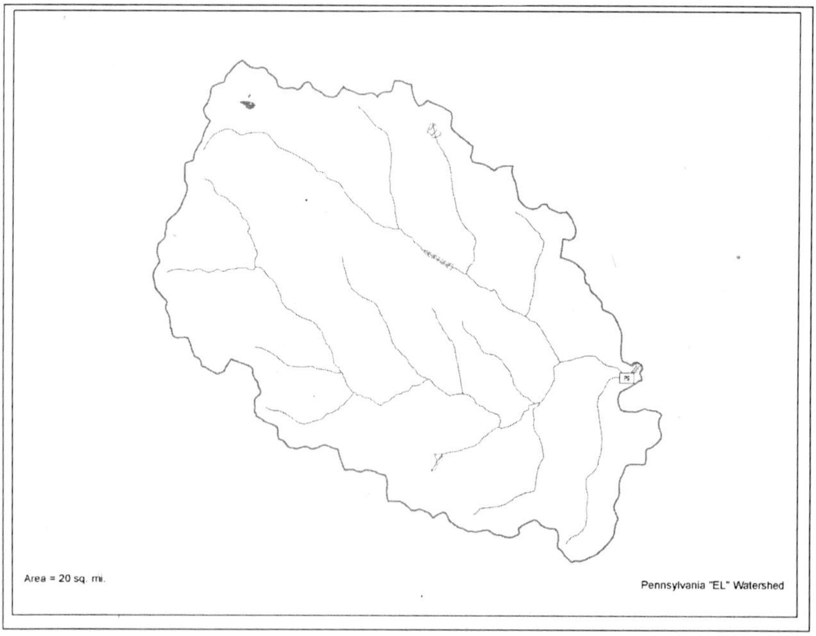

Figure 11. GIS presentation of the Pennsylvania PBA-VN watershed.
(see page 8 of color insert)

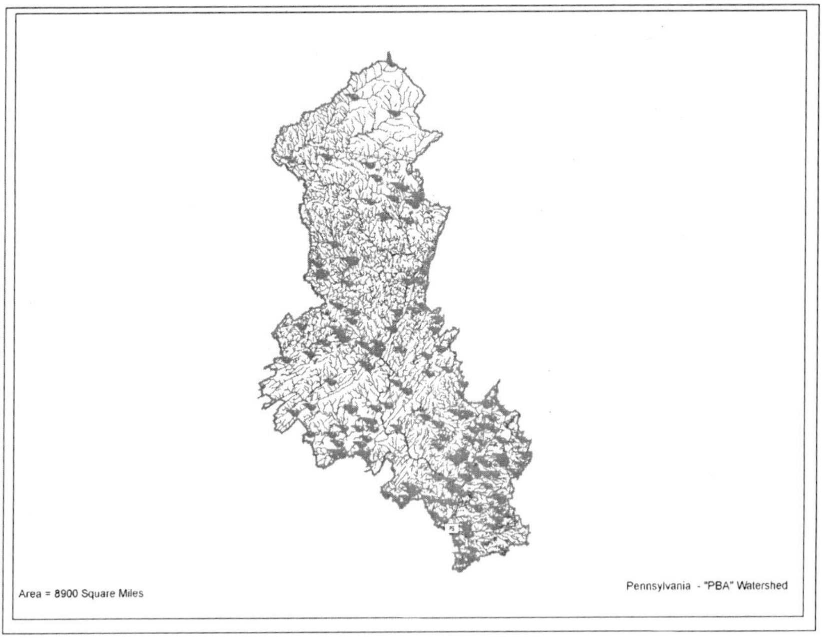

Figure 12. GIS presentation of the Pennsylvania PBE-VN watershed. (see page 8 of color insert)

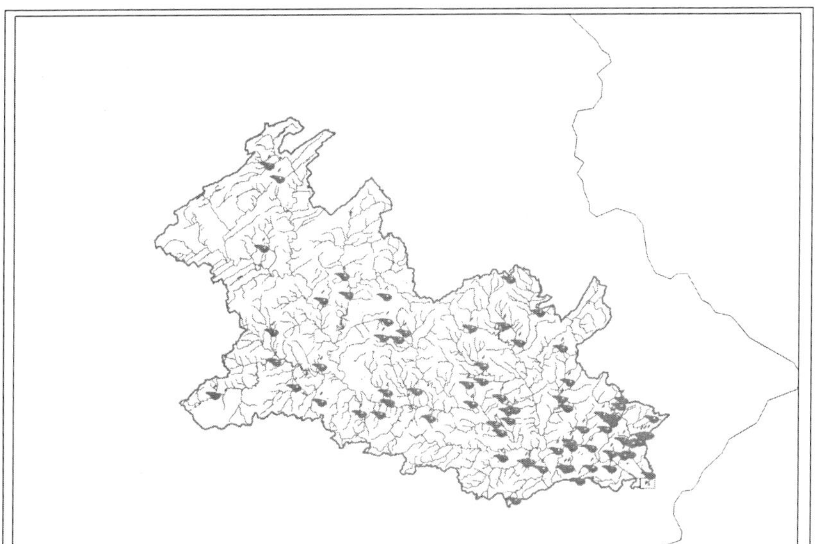

Figure 13. GIS presentation of the Pennsylvania BF-VN watershed. (see page 9 of color insert)

Results and Discussion

Calculation results are presented before results from the monitoring portion of this work. Predictions from the standard turf grass scenario are provided as a series of different time points (e.g. 4, 21, 60 days). However, for the purposes of this work, we report only the acute or instaneous predictions from the models as a worst case exposure concentration. Using the maximum label rate and number of applications as prescribed by USEPA standard methodology, PRZM/EXAMS predicted a maximum concentration of 84.3 ug/L. Therefore, for regulation, 84.3 ug/L would be the concentration expected in community water supplies based on USEPA methodology. Applying the correction method of Jackson et al. (*7*), an anticipated upper bound exposure for the same community water supply would be 0.07 ug/L. The calculation result for the correction factor (Figure 14) was derived as follows (Table III):

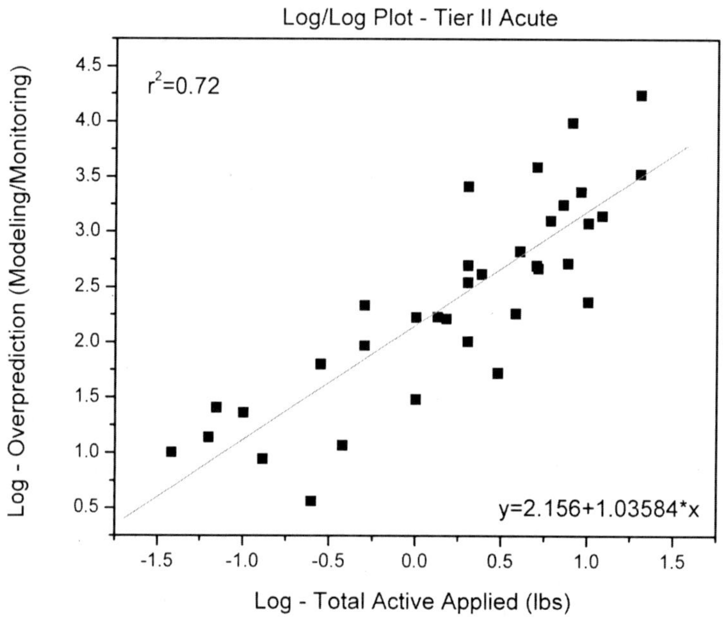

Figure 14. Log/log plot of the relationship between PRZM/EXAMS exposure estimates and model overprediction.
(see page 9 of color insert)

Table III. Derivation of PRZM/EXAMS Correction Factor for Estimation of Vinclozin Exposure of Drinking Waters Supplies

Application Rate (lb/A)	*Application Rate (Log)*	*R*[a]	*OPF*[b]	*Concentration (Predicted) (μg/L)*	*Concentration (Corrected) (μg/L)*
8	0.903	3.09	1234.4	84.3	0.07

[a] Regression
[b] Overprediction Factor

Results from both the BASF and USEPA GIS dilution methods are presented in Figures 15 and 16. The BASF dilution method predicted a concentration range from a low of 0.35 ug/L to a maximum of 2.45 ug/L, depending on the watershed specific inputs. The USEPA dilution method predicted a concentration ranging from 25.36 ug/L to 45.6 ug/L, depending on the watershed specific inputs. Results from the WARP calculations are presented in Figures 17 and 18. WARP-predicted maximum concentrations ranged from 0.17 ug/L in the Pennsylvania watersheds to 1.09 ug/L in the Ohio watersheds, depending on specific inputs. A summary of the various calculation method results is presented in Table IV. Table IV also includes results from the monitoring study, which were all non-detects. From these results, it is possible to draw some conclusions about the exposure estimate methods. Whether we use the USEPA dilution method or the uncorrected PRZM/EXAMS standard turf grass scenario, exposure is greatly overpredicted by the regulatory estimation methods. Interestingly, the mean exposure estimate value from the USEPA dilution method is approximately the same as the PRZM/EXAMS standard turf grass scenario prediction. The WARP method and the BASF dilution method provided approximately the same exposure estimate for the watersheds, while the PRZM/EXAMS standard scenario corrected (*7*) further improved the exposure estimate by an order of magnitude. Figure 19 is a diagram presenting the calculation methods used and the varying levels of prediction refinement provided by each method.

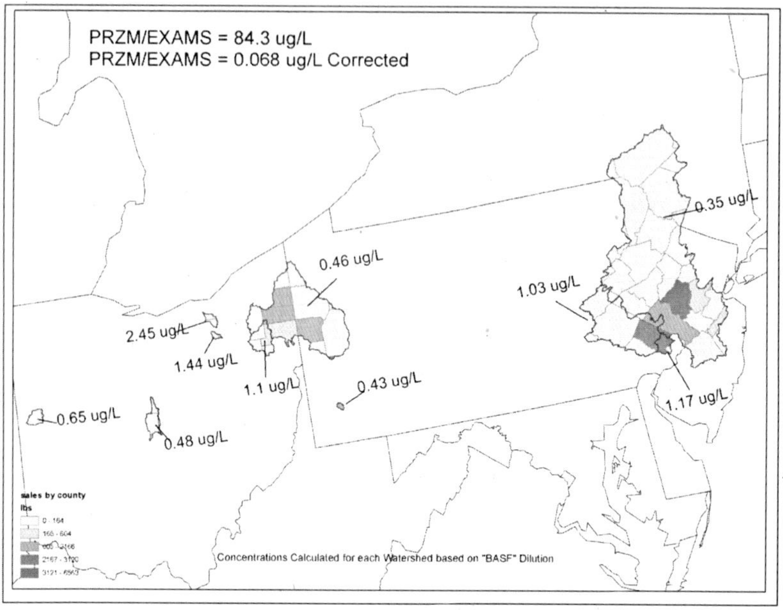

Figure 15. Predicted exposure concentrations in the study watersheds using the BASF dilution calculation.
(see page 10 of color insert)

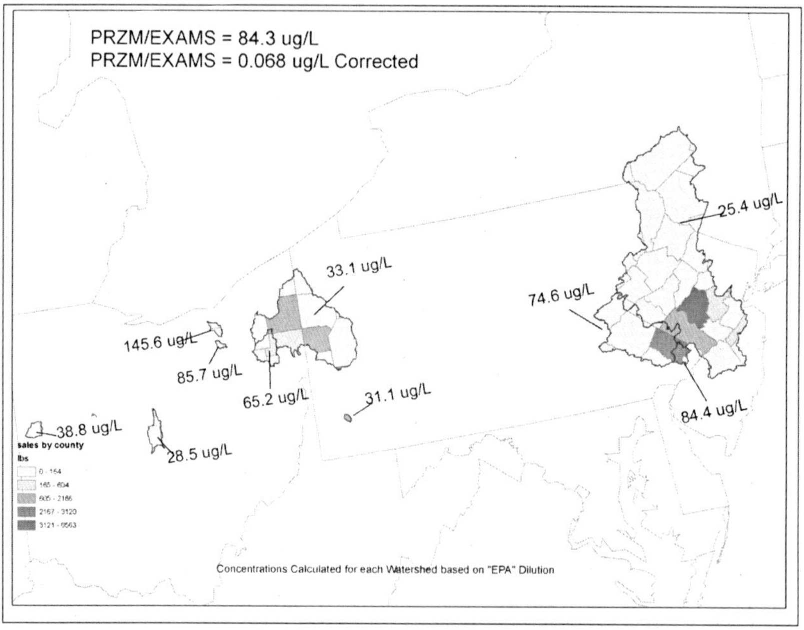

Figure 16. Predicted exposure concentrations in the study watersheds using the USEPA dilution calculation.
(see page 10 of color insert)

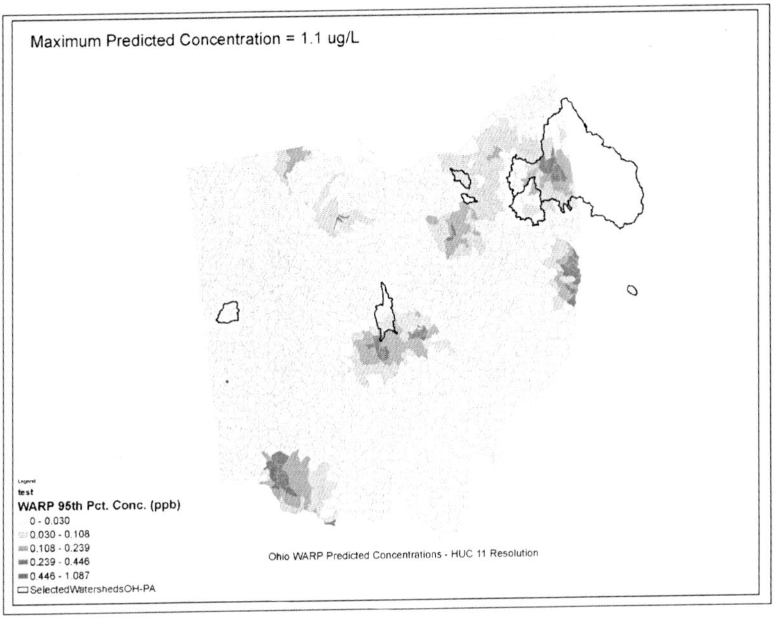

Figure 17. Predicted exposure concentrations in the study watersheds using WARP for the state of Ohio.
(see page 11 of color insert)

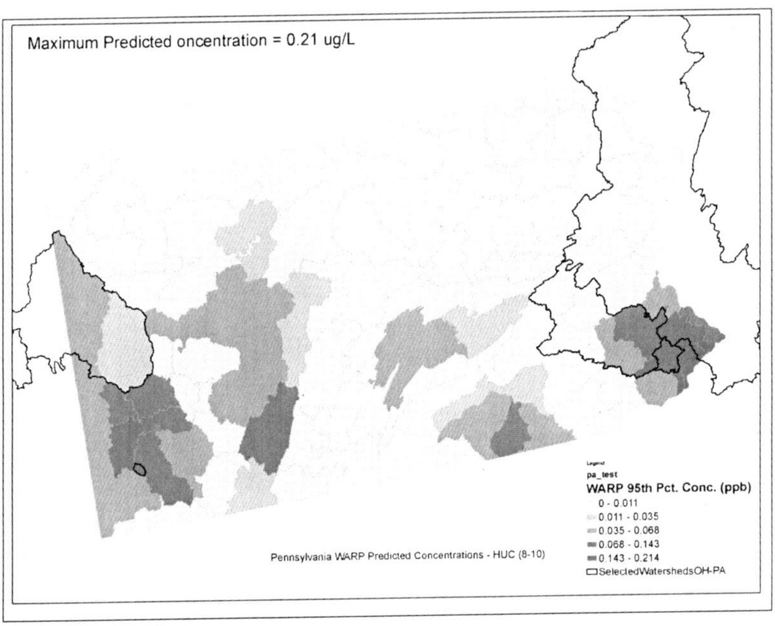

Figure 18. Predicted exposure concentrations in the study watersheds using WARP for the state of Pennsylvania.
(see page 11 of color insert)

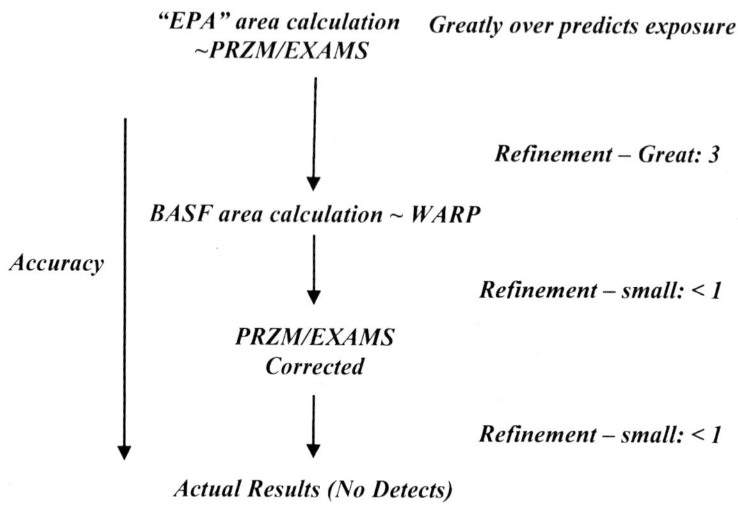

Figure 19. Diagram illustrating the reduction in exposure overprediction moving from specific modeling methods to monitoring. Arrows indicate flow from all directions into the water body.

Table IV. Summary of Watershed Concentrations Based on Various Calculation Methods

System ID	*BASF Calc. (ug/L)*	*"EPA" Calc. (ug/L)*	*WARP (ug/L)*	*PRZM/ EXAMS (ug/L)*	*PRZM/ EXAMS (ug/L) corr.*	*Monitoring (ug/L)*
OH-BA-VN	1.10	65.17	1.09	84.3	0.068	N.D.
OH-BE-VN	1.44	85.68	1.09	84.3	0.068	N.D.
OH-CE-VN	0.48	28.52	1.09	84.3	0.068	N.D.
OH-CH-VN	0.65	38.82	1.09	84.3	0.068	N.D.
OH-NF-VN	2.45	145.59	1.09	84.3	0.068	N.D.
PA-EL-VN	0.43	31.06	0.17	84.3	0.068	N.D.
PA-NO-VN	0.46	33.12	0.17	84.3	0.068	N.D.
PA-PBA-VN	1.03	74.62	0.17	84.3	0.068	N.D.
PA-PBE-VN	0.35	25.36	0.17	84.3	0.068	N.D.
PA-BF-VN	1.17	84.37	0.17	84.3	0.068	N.D.

N.D. = Not Detected

USEPA's Screening Concentration In GROund Water (SCI-GROW) model was also used to estimate ground water exposure for this analysis. The authors view predictions from SCI-GROW as an upper bound exposure estimate. Use data or the application rate entered into the SCI-GROW model was the total seasonal use rate based on the product label. The SCI-GROW model was developed from data generated in regulatory guideline prospective ground water studies (PGWs), which do not provide results reflective of exposure estimates in drinking water wells. Results from the SCI-GROW model predicted exposure in water of approximately 0.93 ug/L, while monitoring results were all non-detects.

Large Magnitude Model Overpredictions

Based on the calculation methods used in this work, it is evident that some of the methods result in large magnitude over predictions. Two of the methods producing the largest magnitude overpredictions were the USEPA standard turf grass scenario method and the USEPA GIS dilution method. Part of the reason for the large magnitude overprediction in these approaches is the assumption about use rate. The USEPA methods take the maximum allowable amount of product that can be used from the product label and use that amount as model input, while the other methods are based on actual sales data. Using a more descriptive application rate with the USEPA methods does move the predicted exposure concentrations in a directionally correct fashion, however it does not address the watershed conceptual problem with the standard turf grass scenario method. Since the PRZM/EXAMS standard scenario methodology is the principal method by which the USEPA Office of Pesticide Programs conducts exposure estimation for drinking water, comments on the conceptual model used as the basis for this methodology are presented specifically to that method. It is important to mention that if an exposure model estimation method is going to be descriptive of an environment (in this case a watershed), the model(s) must properly describe the important aspects of that environment which drive the potential for chemical exposure. Many of the standard scenario methodology issues fail at this point in the process. For example, spray drift from product application enters the standard water body from all directions simultaneously. The model does not account for runoff from untreated areas or areas treated with another product; there is no base flow into the water body, there are no vegetative filter areas or non crop areas which the runoff flows through, nor is there a temporal component between when a runoff event occurs and when the runoff mass reaches the water body (it is currently considered to be instantaneous). A diagrammatic comparison of the USEPA conceptual model to an actual watershed can be found in Figure 20.

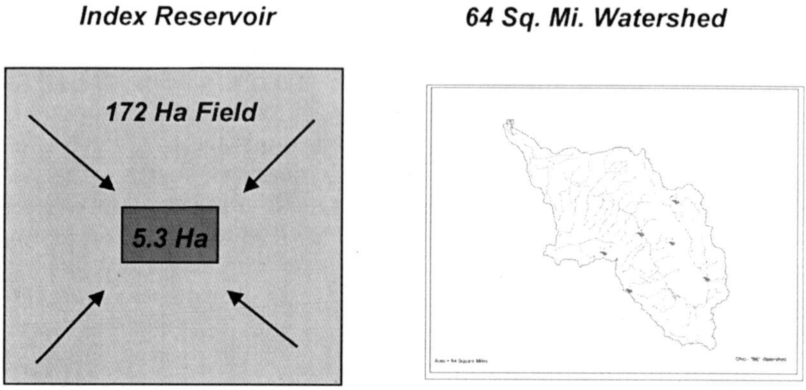

Figure 20. Comparison of the Index Reservoir conceptual model and a GIS coverage of an actual watershed. The arrows indicate aerial drift enters the water body from all directions.
(see page 12 of color insert)

Consequences of Large Magnitude Model OverPredictions

We have described some factors leading to large magnitude exposure over prediction in watersheds. Unfortunately, results from methods producing large magnitude overpredictions are published in public sources such as the federal register on a regular basis. One of the biggest concerns from large magnitude exposure overprediction is public perception. The general public is often familiar with science issues, but generally does not understand them well. For example, it might happen that a local newspaper reports that concentrations in the local drinking water supply are anticipated to be 80 ug/L based on the USEPA methodology, but in truth, using a more realistic exposure estimation method could indicate exposure in the low ng/L concentration range. Further, based on monitoring data, results might indicate that there is in fact no exposure of drinking water supplies (which is commonly the case) from the use of the crop protection product. A second concern from large magnitude model overprediction is the unnecessary filling of the human health dietary risk cup, which in turn limits the allowable development of a given crop protection product in the marketplace. A final concern resulting from large magnitude model overpredictions is the money spent doing investigative studies where it is not necessary. If exposure estimations were properly conducted, it is probable that many studies would not be required and the money, time and effort could be directed into more worthwhile examinations.

Finally, we believe that the use of watershed models (such as WARP) and tools such as GIS have greatly improved our ability to conduct predictive exposure estimates that are still protective in a regulatory sense. Industry strongly recommends adopting a "weight of evidence" approach in conducting exposure estimation based on appropriate modeling, GIS and monitoring data of suitable quality in order to determine probable exposure concentrations.

References

1. Durborow, T.E.; Barnes, N.L.; Cohen, S.Z.; Horst, G.L.; Smith, A.E. 2000. Calibration and Validation of Runoff and Leaching Models for Turf Pesticides, and Comparison with Monitoring Results. Fate and Mgt of Turfgrass Chems, ACS Vol 743, 195-227.
2. Haith, D.A. Turf PQ, A Pesticide Runoff Model for Turf. *J. Environ. Qual.* **2001,** *30*, 1033-1039.
3. Ma, Q.L.; Hook, J.E.; Wauchope, R.D.; Dowler, D.D.; Johnson, A.W.; Davis, J.G.; Gascho, G.J.; Truman, C.C.; Sumner, H.R.; Chandler, L.D.. GLEAMS, Opus, PRZM2B, and PRZM3 Simulations Compared with Measured Atrazine Runoff. *Soil Sci. Soc. Am. J.* **2000**, *64,* 2070-2079.
4. Pearce RA, Trlica MJ, Leninger WC, Smith JL, Frasier GW. Efficiency of Grass Buffer Strips and Vegetation Height on Sediment Filtration in Laboratory Rainfall Simulations. *J. Environ. Qual.* **1997,** *26*, 139-144.
5. Rhodes, R.C., Long, J.D. Runoff and Mobility Studies on Benomyl in Soils and Turf. *Bull. Env. Contam. Toxicol.* **1974,** *12(4)*, 385-393.
6. Wauchope, R.D. Pesticides in Runoff: Measurement, Modelling and Mitigation. *J. Environ. Sci. Health.* **1996,** *B31 (3)*, 337-344.
7. Jackson, S.H., P. Hendley, M. Russell, N. Poletika, and R. Jones. Comparison of Regulatory Method Estimated Drinking Water Exposure Concentrations with Monitoring Results from Surface Drinking Water Supplies, *J. Agri. Food Chem.* **2005,** *53*, 8840-8847.
8. ILSI. *A Framework for Estimating Pesticide Concentrations in Drinking Water for Aggregate Exposure Assessments*; International Life Sciences Institute: Washington, DC, **1999**; http://rsi.ilsi.org/NR/rdonlyres/ACB8DC87-FE21-44CA-8D00-348AFE8FBF2B/0/dwpests.pdf
9. Crawford C. G., S. J. Larson, and R. J. Gilliom. Development and Application of the Watershed Regression For Pesticides (WARP) Methods To Estimate Organophosphate Pesticide Concentration Distributions In Streams. U.S. Geological Survey Water-Resources Investigations Report 2003 Draft, Prepared in cooperation with the U.S. Environmental Protection Agency.
10. Dmrkynetec, St. Louis, MO 63146, https://doanemrwebsc.com.
11. Smith, A.E.; Bridges, D.C. Movement of Certain Herbicides Following Application to Simulated Golf Course Greens and Fairways. *Crop Science* **1996**, *36*, 1439-1445.
12. Carleton, J.N.; Lin, J.; Corbin, M. Development of a Modeling Approach to Estimate Runoff of Pesticide Residues from Managed Turf Grass. United States Environmental Protection Agency, EFED Turf Modeling Workgroup. 2002.
13. United States Environmental Protection Agency . Florida Turf Scenario in Pesticide Root Zone Model Field and Orchard Crop Scenario Metadata. 2003, http://www.epa.gov/oppefed1/models/water/index.htm.
14. United States Environmental Protection Agency. Pennsylvania Turf Scenario in Pesticide Root Zone Model Field and Orchard Crop Scenario Metadata. **2003,** http://www.epa.gov/oppefed1/models/water/index.htm.

Chapter 7

Turfgrass Dissipation of Cyazofamid

Jerome L. Wiedmann

ISK Biosciences Corporation, 7470 Auburn Rd., Suite A, Concord, OH 44077

Traditionally, the fate of pesticides applied to turf has been measured by a 120-day dissipation study to satisfy the United States Environmental Protection Agency (USEPA) data requirements. The focus of these studies has been to measure the rate of degradation of the active ingredient and its degradates as well as to assess the potential for leaching of each of these compounds into the soil. Historically, all grass clippings were left on the plot to recycle the residues. Recently, concern has also been raised about the loss of residues from the treated area via removal of the grass clippings during the mowing process. The fate of cyazofamid, a cyanoimidazole fungicide, used for *Pythium* control in turf was investigated with and without removal of clippings at sites in North Carolina and Virginia. Comparison of loss of cyazofamid by clipping removal to other mechanisms of dissipation indicates that for a pesticide such as cyazofamid, which has very short foliar and soil half-lives and which does not leach, the loss due to removal in the clippings is minimal (0.7-2.2%). Even with application rates of 1 lb a.i./A, dissipation of cyazofamid from turf was rapid (half-lives of 18 to 19 days). There was very little movement of either cyazofamid or its degradates into the soil. By 120 days after the last of three applications, 94-98.8% of the applied cyazofamid had degraded (primarily in the form of bound residues).

Background

Cyazofamid [4-chloro-2-cyano-*N*,*N*-dimethyl-5-*p*-tolylimidazole-1-sulfonamide (IUPAC), 4-chloro-2-cyano-*N*,*N*-dimethyl-5-(4-methylphenyl)-1*H*-imidazole-1-sulfonamide (CA)] is a fungicide developed by Ishihara Sangyo Kaisha, Ltd. It has excellent activity on all life stages of Oomycete fungi such as *Phytophthora, Plasmopara, Pythium, Pseudoperonospora* and *Aphanomyces*. Effective preventative activity is observed at 60-80 g/ha (0.052-0.071 lb a.i./A). Cyazofamid is currently registered in the United States for use on potatoes, tomatoes, cucurbits and imported wine grapes. Additional uses are under investigation. *Pythium* control on turf grass is a pending registration.

Important properties of cyazofamid include low water solubility (0.107 mg/L) and a low vapor pressure ($<1.33 \times 10^{-5}$ Pa) (*1*), as well as a lack of ionization between pH 2 to 12 and a log K_{ow} of 3.2 (*2*). Even after multiple spray applications, the half-life in either plants or soil is $\leq$3 days for multiple applications of 0.071 to 0.089 lb a.i./A (*2*). Currently labeled use rates are 6-10 applications of 0.071 lb a.i./A applied at 7-day intervals. Usage for *Pythium* control on turf requires a higher application rate, potentially as high as 1 lb a.i./A, with up to three applications at 7-day intervals. Thus, for use on turf, the total seasonal application of cyazofamid may increase by a factor of four while the time between the first and last applications drops by a factor of 3 to 4.5, as compared to crop uses.

Previous soil dissipation studies with cyazofamid were conducted on bareground plots. Turf is covered with a thatch layer and a dense grass population that may influence the rate of dissipation of cyazofamid and its degradates. Consequently, dissipation studies in turf were needed to determine the rate of decline of cyazofamid under these conditions and to check for possible leaching. In addition, the literature on the dissipation of fungicides in turfgrass is very limited (*3*).

Study Design

The study design for turf soil dissipation studies is clearly defined in USEPA OPP guidelines (*4*). Consultation with R. David Jones of USEPA (*5*) revealed that modification of the study design would be needed since these studies are not currently focused on leaching but rather on major routes of dissipation. The traditional endpoints measured in this study were movement of the pesticide into the soil and rate of degradation; clippings from mowing the grass were left on the plots. A new requirement was measurement of the loss of residue that could occur from removal of the clippings.

The traditional study contained a control plot and a treated plot divided into 3 subplots, so that treated samples were obtained in triplicate. Five cores were taken per subplot at each time point and composited. The maximum use rate (three weekly applications of 1 lb a.i./A) was used. Sampling intervals were –1 and 0 days after application 1, –1 and 0 days after application 2, –1 and 0, 1, 2, 4, 7, 14, 21, 28, 60, 90 and 120 days after application 3. Both the grass/thatch layer and the soil beneath were sampled and analyzed. The soil samples were divided into three-inch increments and analyzed at increasing depths until residues were non-detectable.

In addition to the traditional objectives, determination of the loss due to clipping removal required significant study modification. Determination of the potential loss of residues due to clipping removal could not be easily done in the context of the standard experiment, since clippings needed to be left in place for the traditional study (to determine total residue) and needed to be removed for the new study (to measure residues in clippings). In addition, the traditional study had clearly defined sampling times based on expected decline of the active ingredient, however, the sampling times for the new study were dependent on the rate of growth of the grass. Both of these problems were resolved by establishing two treated plots, one for the traditional study and a second to measure the clipping removal effect. The two plots needed to be identical initially for comparison purposes. In both studies, the plots were end to end and treated in the same spray pass. Plot 2 (the control plot was Plot 1) was sampled deep enough to measure the downward movement of cyazofamid in soil. Three-foot cores were used in this case, since minimal movement was predicted based on results from bare ground dissipation studies. Plot 3 was included to determine the effect of removal of clippings, thus movement through soil was a less important parameter, so only 0-6 inch depth cores were sampled for comparison to the clippings and for comparison to the other plot. A set of five 2.5-3 inch diameter cores provided ample sample weight for analysis of soil and grass/thatch. It did not provide adequate sample for clipping analysis; it was therefore necessary to go to a 1 foot square frame as deep as the mower setting. All of the grass higher than this frame was clipped for the sample, and even with 5 square feet of sampling area, an occasional sample of clippings weighed less than 5 grams. After the clippings were removed, the cores were taken from the center of each clipped area. After each sampling, the whole plot was mowed with a mower that collected the clippings. Plot 2 was mowed with a mulching mower on the same schedule to keep all clippings on the treated plot.

Field Site Information

Two sites were selected, North Carolina (USEPA Region 2) and Virginia (USEPA Region 1). The North Carolina study was conducted at the former Aventis Cropscience Research Facility near Pikeville, NC. The Virginia study was conducted at the Brookmeade Sod Farm, Inc. near Ashland, VA. The plots were very different in appearance due to site management practices. North Carolina had a thin stand of fescue grass, variety Kentucky 31, at the initial application since it had grown out and been mowed short just prior to study initiation. The Virginia site was on a sod farm with a thick stand of a blend of 90% Tall Fescue (equal parts of Laramie, Stetson and Bravo cultivars) with 10% Shamrock Kentucky Bluegrass. At NC, a 15 foot boom was used and the spray applied in two passes. Sampling was random in the direction of spraying, and non-random across the spray pattern, with 5 sections of each sub-plot sampled. At VA, a 5 foot boom was used with each subplot receiving one spray pass. Sampling was from one random five-foot square section of each sub-plot, with all 5 cores pulled from that section in an X pattern. At NC, sampling was with a Giddings hydraulic soil probe. At VA, sampling was with a jackhammer driven probe and the cores were removed with a foot jack.

Analytical Methodology

The degradation of cyazofamid in soil and plants is well defined (*2*). In an initial step, the sulfonamide group is lost to form CCIM, [4-chloro-5-*p*-tolylimidazole-2-carbonitrile (IUPAC), 4-chloro-5-(4-methylphenyl)-1*H*-imidazole-2-carbonitrile (CA)]. From CCIM two pathways occur. The methyl group may undergo oxidation to the acid, forming CCBA, [4-(4-chloro-2-cyanoimidazol-5-yl)benzoic acid (IUPAC), 4-(4-chloro-2-cyano-1*H*-imidazol-5-yl)benzoic acid (CA)]. The alternate pathway involves conversion of the cyano group to the amide CCIM-AM, [4-chloro-5-*p*-tolylimidazole-2-carboxamide (IUPAC), 4-chloro-5-(4-methylphenyl)-1*H*-imidazole-2-carboxamide (CA)] which than is further converted to the acid, CTCA, [4-chloro-5-*p*-tolylimidazole-2-carboxylic acid (IUPAC), 4-chloro-5-(4-methylphenyl)-1*H*-imidazole-2-carboxylic acid (CA)]. CTCA and CCBA, after formation, bind to the soil organic matter, yielding high percentages of bound residues in radioactive studies (*2*). Figure 1 illustrates this process.

For soil, extraction of 50 g of soil was performed using two 100 mL volumes of acetonitrile:water, 3:1 + 0.5% formic acid. Dilution of the extracts to 250 mL with water gave a final solvent mix of 6:4 + 0.4% formic acid. The LC column was a YMC-ODS-AQ S-5 (120 Å 3.0x50mm). LC-MS/MS retention times were 2.0-3.8 minutes when using a 7:3 to 3:7 acetonitrile:water HPLC gradient for 1 to 2.5 minutes. Cyazofamid was run in the negative ion mode and the metabolites were run in the positive ion mode. The thatch method was the same as above except a smaller sample size of 20 g was used, and the first extraction did not contain the formic acid modifier. Both extractions used 40 mL of solvent with dilution to 100 mL. The clippings required further modification of the extraction step. Sample size for clippings was 5 g. The initial extraction used 50 mL of straight acetonitrile. The second extraction used 50 mL of acetonitrile:water (3:1) + 4% formic acid. The limit of detection was consistently 1 ppb in all matrices. The limit of quantitation (LOQ) was consistently 5 ppb with two exceptions. In the clippings matrix interferences were encountered with analysis for CTCA, requiring the LOQ be raised to 50 ppb. Problems with ion suppression were encountered with analysis of CCBA in clippings. This problem was resolved by dilution, which required raising the LOQ for CCBA to 50 ppb in clippings. Since CCBA and CTCA were minor degradates in the clippings which had a minor amount of the total residue these higher LOQ's had no impact on the study.

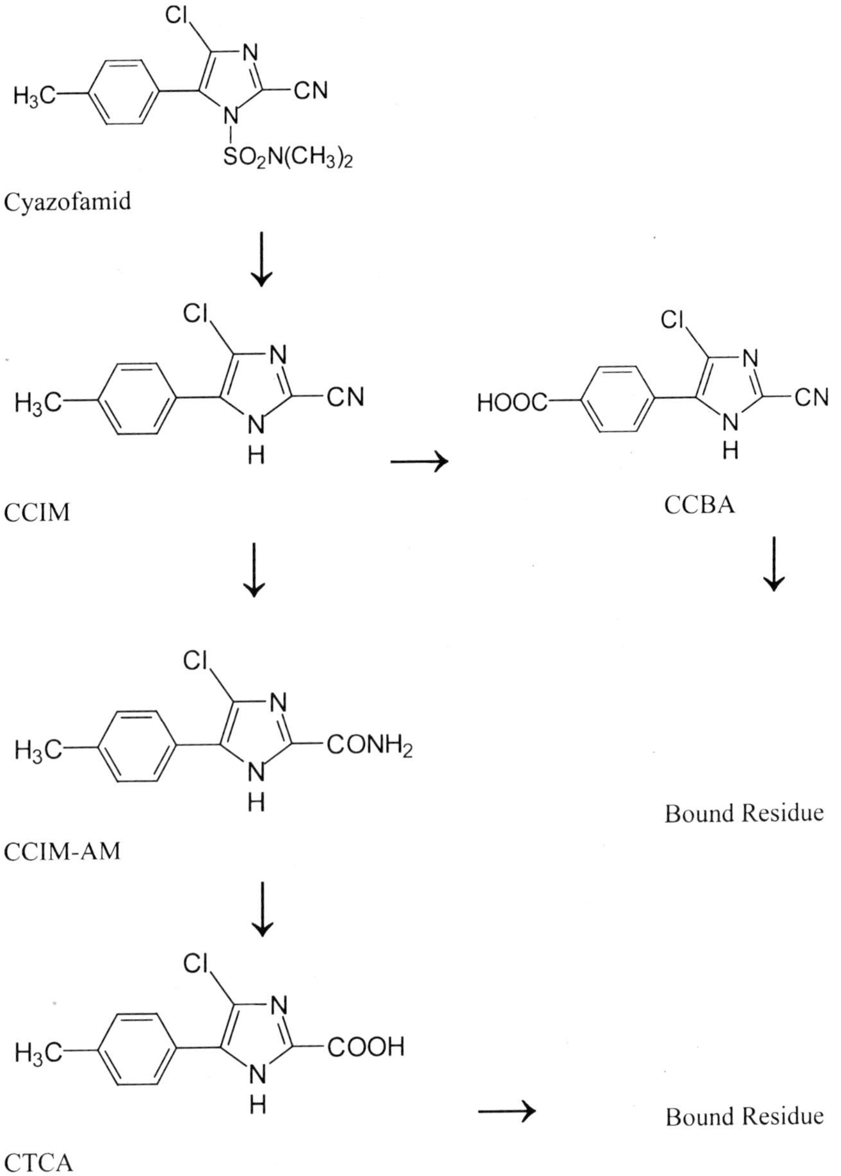

Figure 1. Cyazofamid degradation pathway.

Results

Across the 134 days of each study, the ratios of cyazofamid and its degradates remained relatively constant with no more than 10% of the residue corresponding to any single degradate at any time. Table I shows the consistency of the residue levels across each study.

Table I. Relative Amounts of Cyazofamid and Its Degradates

Compound	*North Carolina* *Percent of Daily Total*	*Virginia* *Percent of Daily Total*
Cyazofamid	80-91	73-96
CCIM	7-8	1-10
CCIM-AM	0-2	0-4
CTCA	0-5	0-9
CCBA	4-8	1-10

Soil cores were split into a grass/thatch layer and three-inch soil increments for analysis. The figures plotted show total residues, however, residues in the soil averaged only 1.7% (NC) to 3.8% (VA) of the total residue for the first 42 days, so the figures are really showing dissipation from the grass/turf layer. Figure 2 shows the decline of cyazofamid at the North Carolina site. Figures 3 through 6 show the declines of CCIM, CCIM-AM, CTCA and CCBA, respectively at North Carolina. Figure 7 shows the cyazofamid decline data for Virginia. Figures 8 through 11 show the declines of CCIM, CCIM-AM, CTCA and CCBA, respectively at Virginia. The residues accumulated as long as applications were being made, dropped off rapidly for the next 3 weeks and then declined at a slower rate. Concentrations of all 4 degradates peaked shortly after the last application and then declined rapidly. CTCA degraded more slowly and consequently built up relative to the other degradates by the end of the study. Excellent agreement between the two sites was observed, even though the plot and sampling designs were significantly different.

For three applications, rainfall events of 0.25 to 0.30 inch occurred during the sampling at 2 hours (VA App. 2) or within a day of treatment (NC App. 2 and VA App. 3). For the other three applications, rainfall events of 0.83 to 1.00 inch occurred 3-4 days after the applications (NC App. 1 and App. 3, VA App. 1). There was no evidence of either increased dissipation or significant leaching to the top soil layer from any of these initial rainfall events. North Carolina received 4.15 inches of rain in the 17 days following the first application. Virginia received 2.81 inches in the same time period. Even with this much rainfall, most of the residue remained in the turf/thatch layer, with only 0-3% of the residue moving down into the soil in North Carolina and 0-1% at Virginia. The highest residues in soil for the North Carolina location generally occurred immediately after application, and thus were likely due to inadvertent transfer of residue from turf/thatch to soil during the coring operation. Cyazofamid is very water insoluble, so lack of downward movement was expected.

Half-life calculations made by regression analysis on the data after the third application indicated a half-life of 19 days at North Carolina, and of 18 days at Virginia. These numbers were consistent with a total application of 3 lb a.i./A over a 2-week period when compared to the bare ground studies, which had total applications of 0.89 lb a.i./A over an eight-week interval. At North Carolina, only 6% of the total applied represented identifiable compounds at 120 days after the last application. Virginia retained only 1.2% over the same period. Thus, even at these high rates, little movement into the soil was observed. Those few residues which did move into the top soil increment were rapidly metabolized and did not move deeper into the soil. The residues collected in the clippings, after suitable mathematical conversions were applied to adjust for the different surface areas collected (clipping residue x area for 5 cores/5 square feet = corrected clipping residue), were surprisingly low, with 2.2% of the applied material removed from North Carolina and only 0.7% removed at Virginia. Losses due to clipping compare well with literature values on other compounds (*6, 7, 8*) where liquid formulations were <1% and granular formulations were 1.2-8%. The formulation used (Ranman® 400SC) is a suspension concentrate which may behave more like a granular than a liquid. This amount of loss in the clippings did not have a significant effect on the rate of dissipation of cyazofamid.

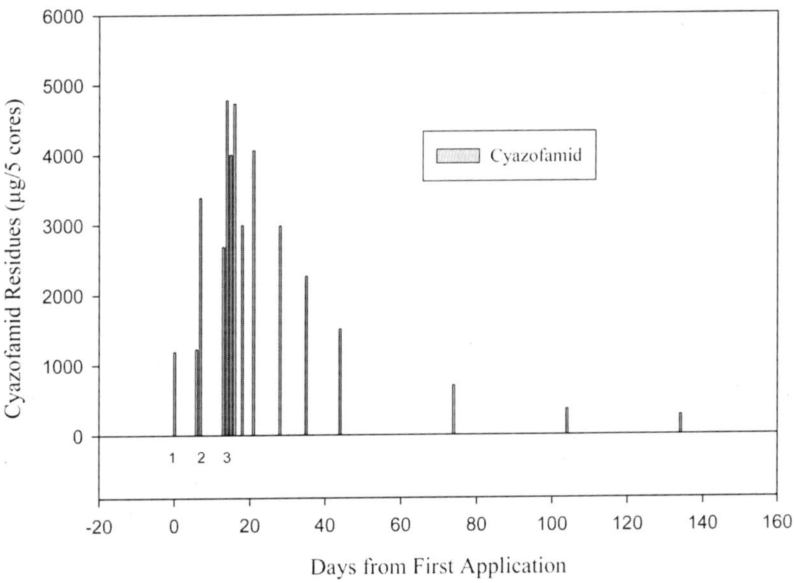

Figure 2. Decline of cyazofamid at North Carolina.

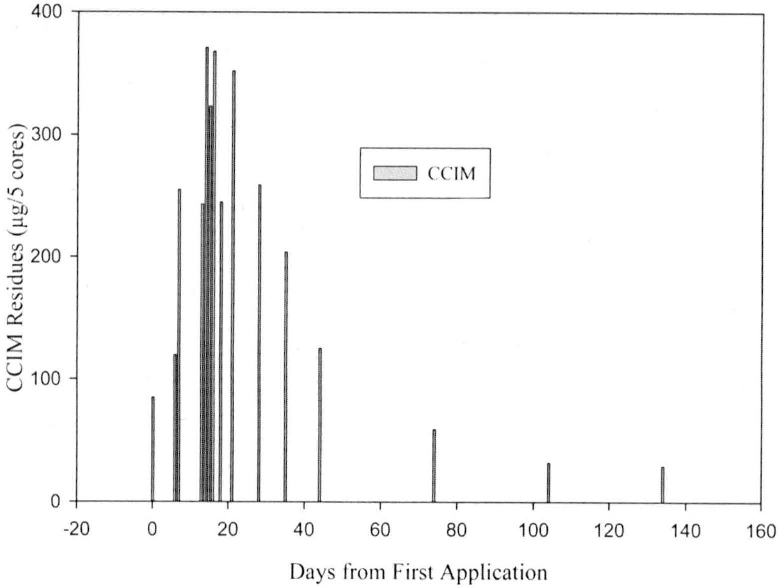

Figure 3. Formation and decline of CCIM at North Carolina.

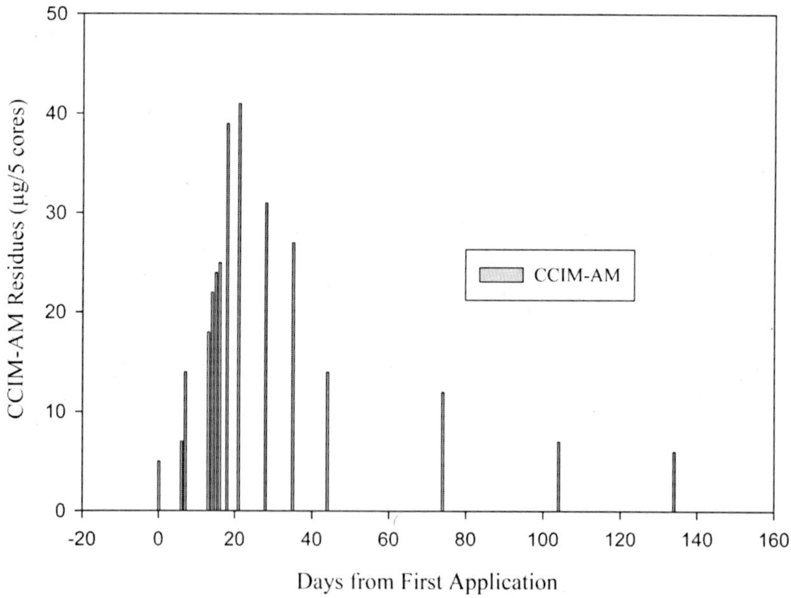

Figure 4. Formation and decline of CCIM-AM at North Carolina.

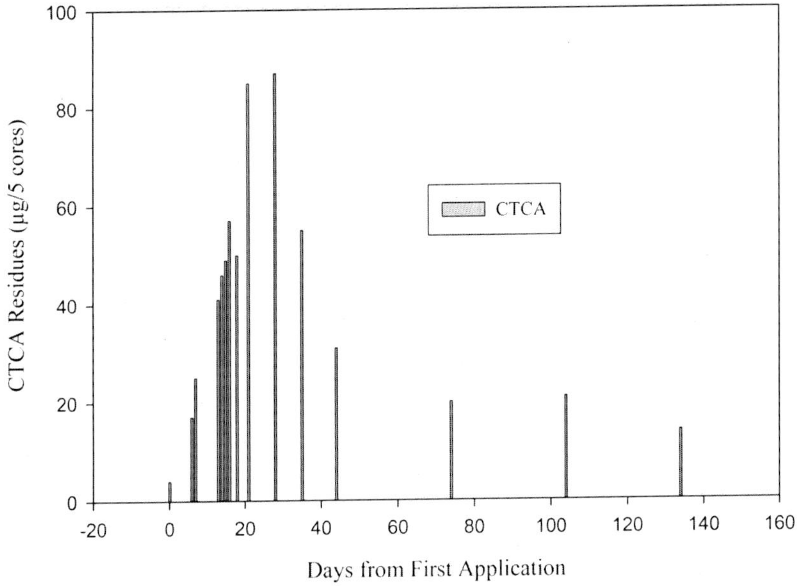

Figure 5. Formation and decline of CTCA at North Carolina.

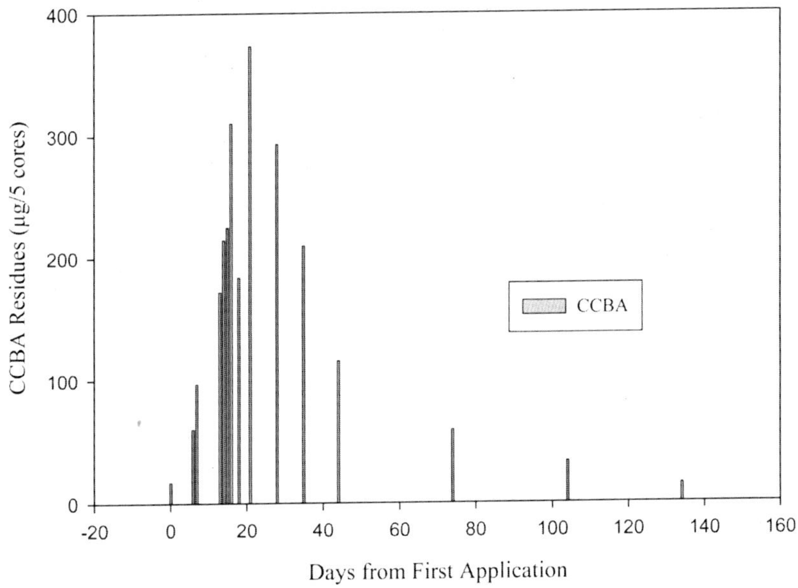

Figure 6. Formation and decline of CCBA at North Carolina.

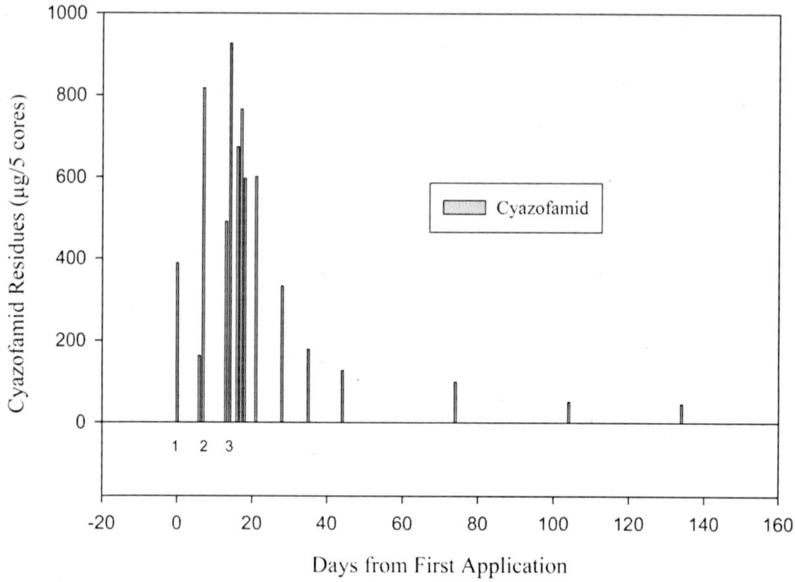

Figure 7. Decline of cyazofamid at Virginia.

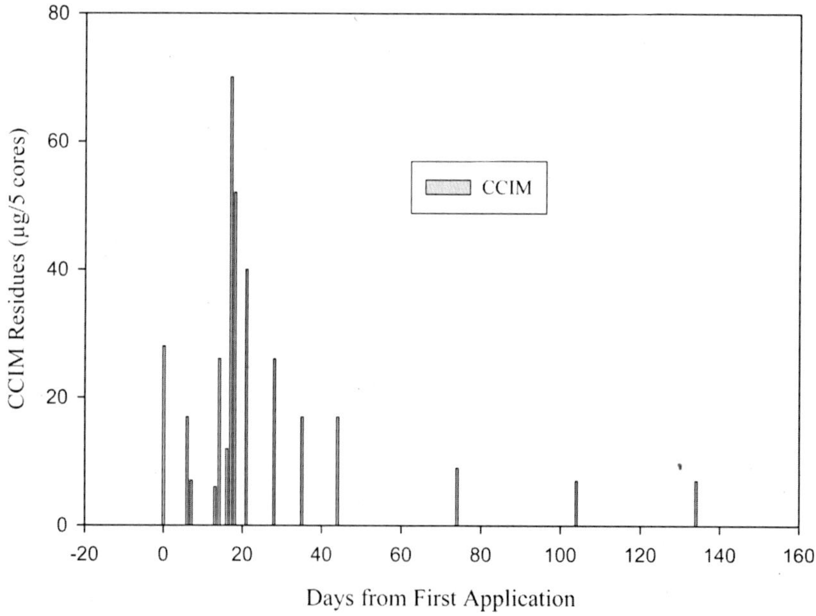

Figure 8. Formation and decline of CCIM at Virginia.

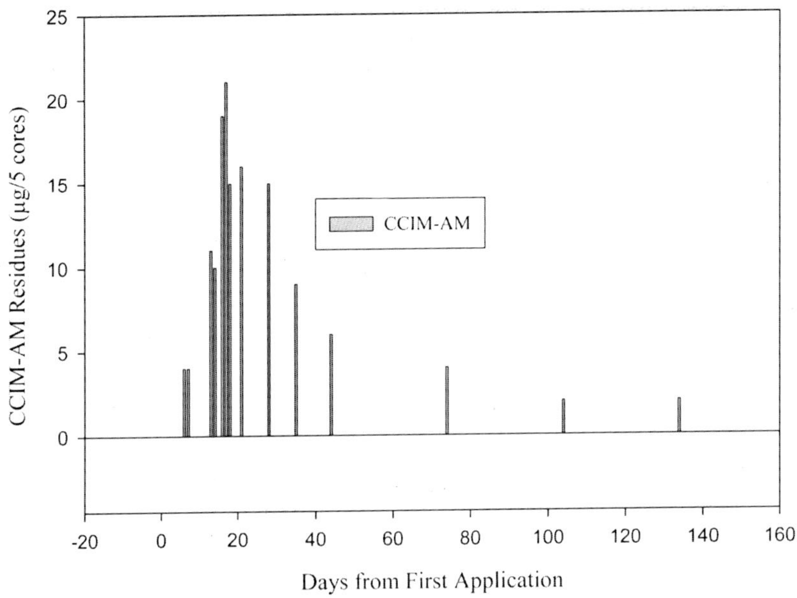

Figure 9. Formation and decline of CCIM-AM at Virginia.

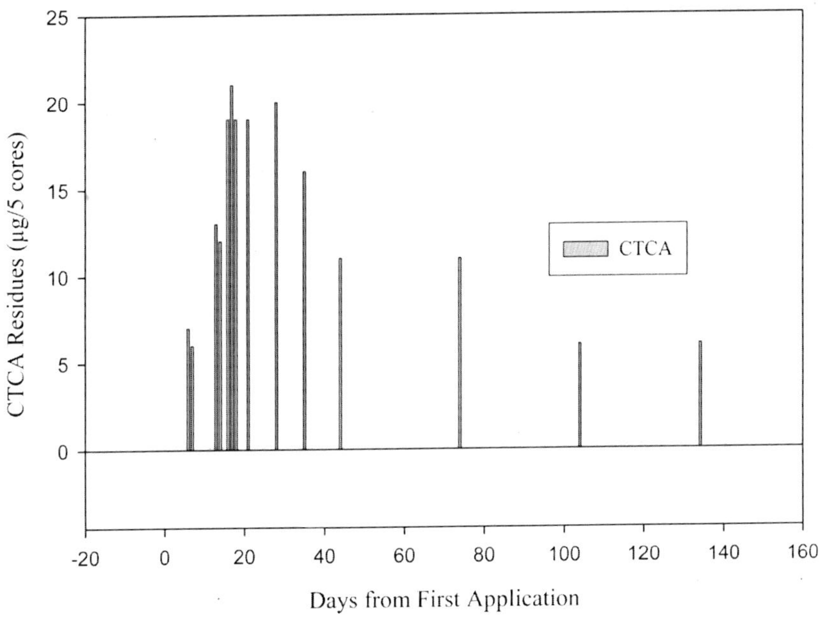

Figure 10: Formation and decline of CTCA at Virginia.

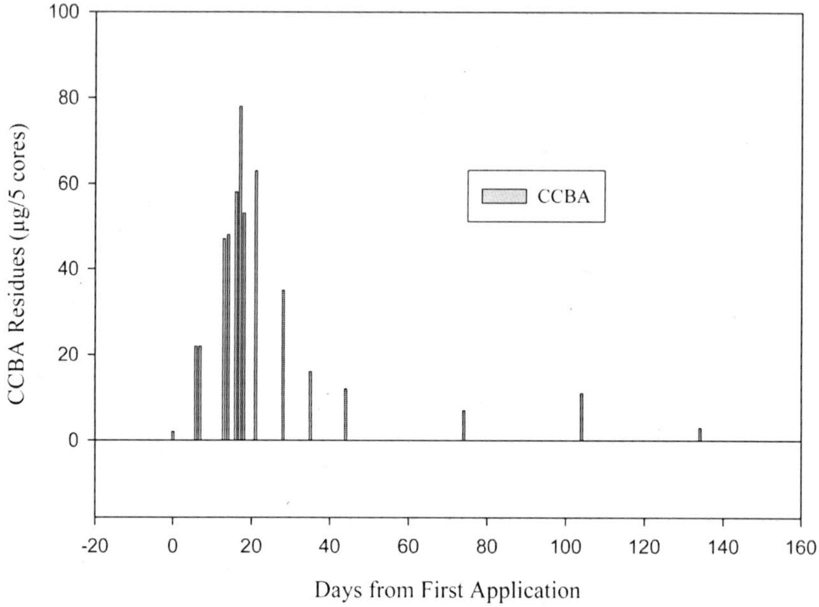

Figure 11: Formation and decline of CCBA at Virginia.

Conclusions

Cyazofamid, applied as three weekly 1 lb/A applications to turf, dissipated rapidly, with half-lives of 18 days at Virginia (correlation 0.9907) and 19 days at North Carolina (correlation 0.9708) using a 2 compartment model. Only 1.2 (VA) to 6% (NC) of the applied cyazofamid remained as cyazofamid plus identifiable degradates at 120 days after the last application. The degradation rates of cyazofamid and its primary degradates were approximately the same, and no degradate became a major part of the measurable residue. Very little movement of cyazofamid and its degradates was observed, either from the grass/thatch layer to the top soil layer, or from the top soil layer to the next deepest soil layer. The amount of cyazofamid and its degradates removed in the clippings was a negligible 0.7-2.2% of the total residues at VA and NC, respectively. The rapid dissipation of cyazofamid in turf minimizes any potential impacts on human or environmental safety.

Acknowledgements

The field portion of the North Carolina study was managed by Dr. Paul V. Garvey Jr. of Agricultural Systems Associates, 3341 Hwy. 58 South, Kinston, North Carolina 28504. The field portion of the Virginia study was managed by James E. Ashley Jr. of Ashgrow Crop Management Systems, Inc., 11913 Simsbury Place, Glen Allen, VA 23059. The analytical portion of the study was conducted by Pyxant Labs Inc., 4720 Forge Rd., Suite 108, Colorado Springs, CO 80907 under the direction of Dr. David A. Robaugh.

References

1. *Material Safety Data Sheet for Technical Cyazofamid*; ISK Biosciences Corporation, 2005.
2. ISK Biosciences Corporation, Unpublished Data.
3. Sigler, W. V., C. P. Taylor, C. S. Throssell, M. Bischoff and R.F. Turco. In Clark, J. M. and Kenna, M. P. (ed.) *Fate and Management of Turfgrass Chemicals*. ACS Symposium Series No. 743. American Chemical Society, Washington, D.C., 2000; pp. 127-149.
4. *Pesticide Assessment Guidelines. Subdivision N, Chemistry: Environmental Fate. EPA-540/9-82-021.* U.S. EPA Office of Pesticide Programs, Washington, DC, 1982.
5. Jones, R. D. USEPA, Private Communication, June 2003.
6. Cisar, J.L. and G.H. Snyder. *Crop Sci.* **1996**, 36: 1433-1438.
7. Wu, L., G. Liu, M.V. Yates, R.L. Green, and P. Pacheco. *Pest Manag. Sci.* **1998**, 58: 335-342.
8. Wu, L., R.L. Green, G. Liu, M.V. Yates, P. Pacheco, J. Gan and S.R. Yates. *J. Environ. Qual.* **2002**, 31:889-895.

Chapter 8

Field Design and Quality Control Considerations for Turfgrass Runoff Studies Conducted for Modeling Purposes

Joseph H. Massey[1], Peter A. Y. Ampim[1], Barry R. Stewart[1], Mark J. Carroll[2], and M. Cade Smith[1]

[1]Department of Plant and Soil Sciences, Mississippi State University, Starkville, MS 39762
[2]Department of Plant Science and Landscape Architecture, University of Maryland, College Park, MD 20742

This paper presents quality control considerations for plot- and field-scale turfgrass runoff studies conducted for modeling calibration and verification. Items addressed include plot construction and maintenance, pesticide application verification, simulated rainfall rate verification, use of conservative tracers to track water movement, soil, thatch layer and weather data collection, and sample handling and storage.

Surface runoff is one of the largest loss mechanisms for pesticides applied to turfgrasses (*1, 2*). Owing to the importance of turfgrass to urban environments and the need to protect water quality, there exists an on-going need to perform turf runoff experiments to (a) assess the behaviors of new chemicals or products, (b) refine best management practices, and (c) calibrate/validate runoff prediction models for turfgrass. Field studies indicate that surface runoff from creeping bentgrass (*3*) and bermudagrass (*4*) is 'scalable' across a range of plot areas. Thus, there is scientific justification for using plot-scale experiments to study the surface runoff of turf chemicals.

Conceptually, conducting a turf runoff experiment is simple: A chemical(s) is applied to grass, subjected to simulated and/or natural rainfall, and runoff collected and analyzed for the test chemical(s). In practice, a runoff study involves a number of steps that must be carefully performed to ensure that scientifically valid, representative data are produced. Small oversights in study design or conduct may compromise data from a scientific or regulatory perspective. The goal of this paper is to assist researchers, and perhaps those charged with evaluating and interpreting runoff study designs and results, by highlighting certain quality control considerations important to the conduct of a plot- or field-scale runoff experiment. This paper is not comprehensive, but presents approaches to quality control that have proven helpful in studies we have conducted over the past four years. For more comprehensive reviews of runoff studies, the reader is directed to Wauchope et al. (*5*) and Nett and Hendley (*6*). The specific aspects addressed here are plot construction and maintenance, pesticide application verification, simulated rainfall rate verification, use of conservative tracers to track water movement, soil, thatch and weather data collection, and sample handling and storage.

Study Planning

A detailed protocol that addresses all aspects of study conduct is critical to the success of studies of this size and complexity. Moreover, an approved protocol is generally required for a study to be submitted to support pesticide registration. A well-designed protocol serves as an invaluable reference throughout a study, as most plot construction and study activities build one upon another.

A thorough literature review is an appropriate place to begin any study of this scale. Unfortunately, quality control programs are not always well defined or explicitly reported in published works. Consultation with chemical manufacturers, experienced researchers, and end-users of the information generated by the study can help in addressing important aspects of study design.

There are certain study details that should not be left to chance, nor addressed as an afterthought once the study is underway. Particular attention should be paid to the methods used to control and account for water movement within the test plots, and those used to account for pesticide application and fate in the turf system. Some pesticides present special considerations, such as those with a propensity to strongly adsorb to surfaces (water solubility $\leq$ 1 mg/L at 25°C), rapidly degraded (soil $T_{1/2} \leq 2$ days), or those that are relatively volatile (vapor pressure > 10^{-4} mm Hg at 25°C). Thus, the researcher must take into account the properties and environmental behavior of the pesticide during protocol development. Sample handling and storage practices are also critical and may be compound dependent. Ultimately, a guiding practice in study design and conduct is to strive to account for as much as possible of the applied rainwater and chemical(s).

Turf Plot Construction and Maintenance

The runoff plot must be constructed so as to capture no more and no less than the actual runoff occurring from the treated plot. Water external to the plot

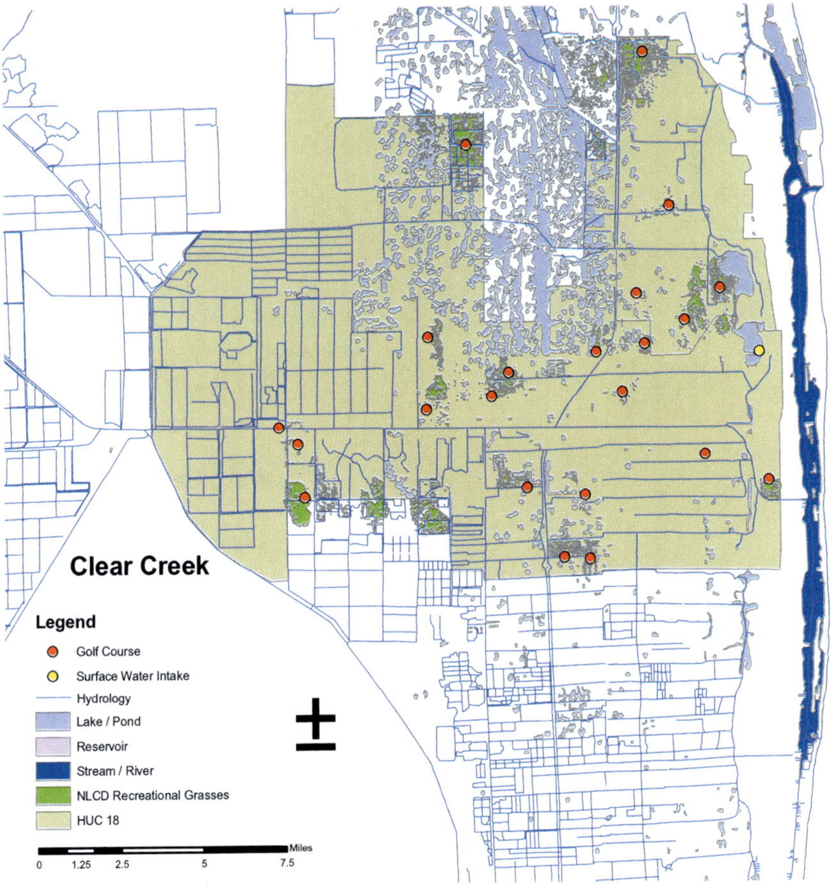

Figure 5.1. Watershed map of West Palm Beach Community Water System

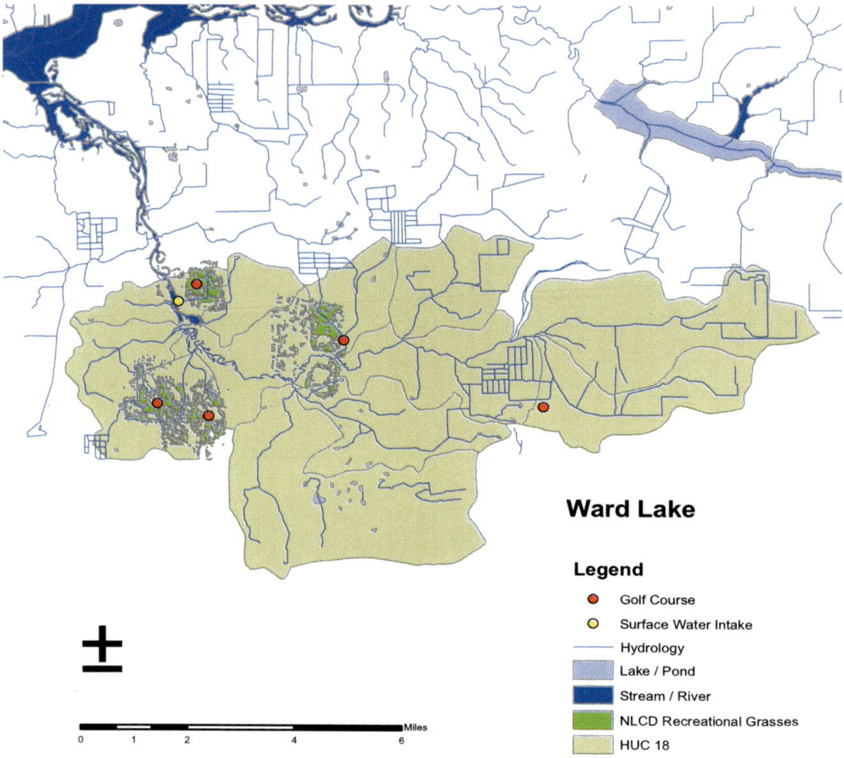

Figure 5.2. Watershed map of Bradenton Community Water System

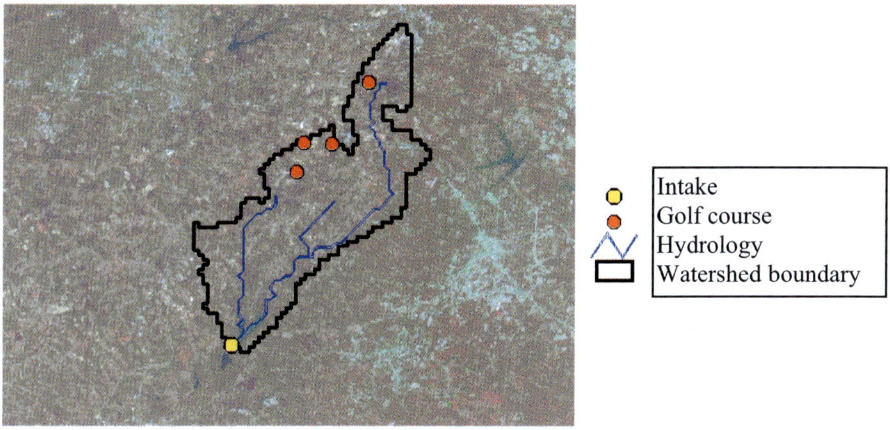

Figure 5.3. Watershed map of Thomasville Community Water System.

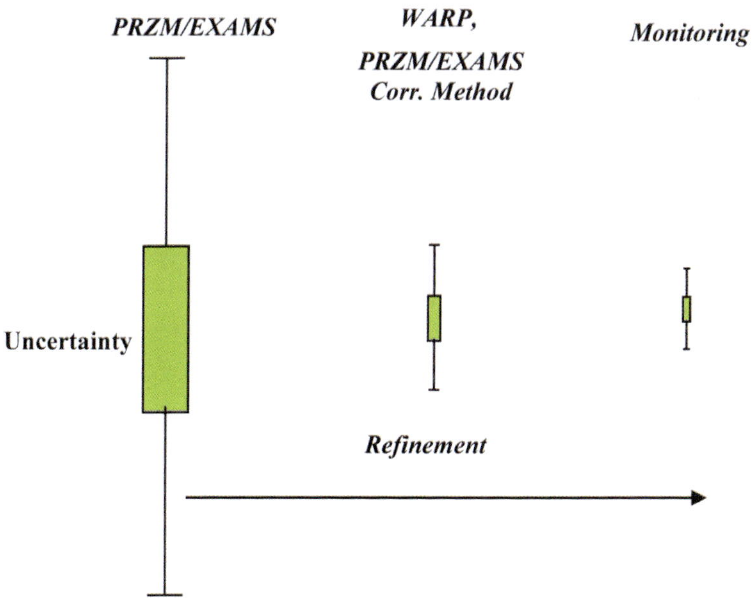

Figure 6.1. Diagram illustrating the reduction in variability moving from modeling to monitoring.

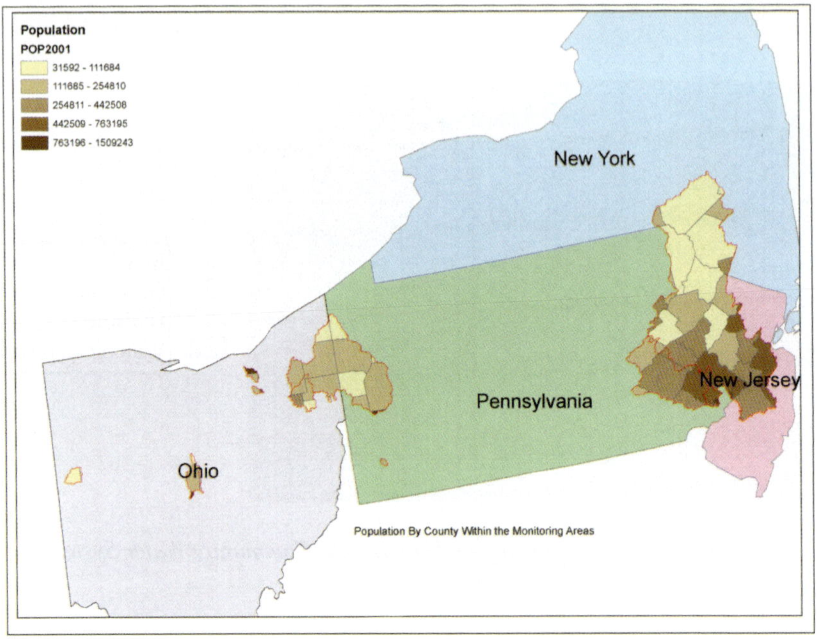

Figure 6.2. General location of the study watersheds within each state. Population within each watershed is symbolized.

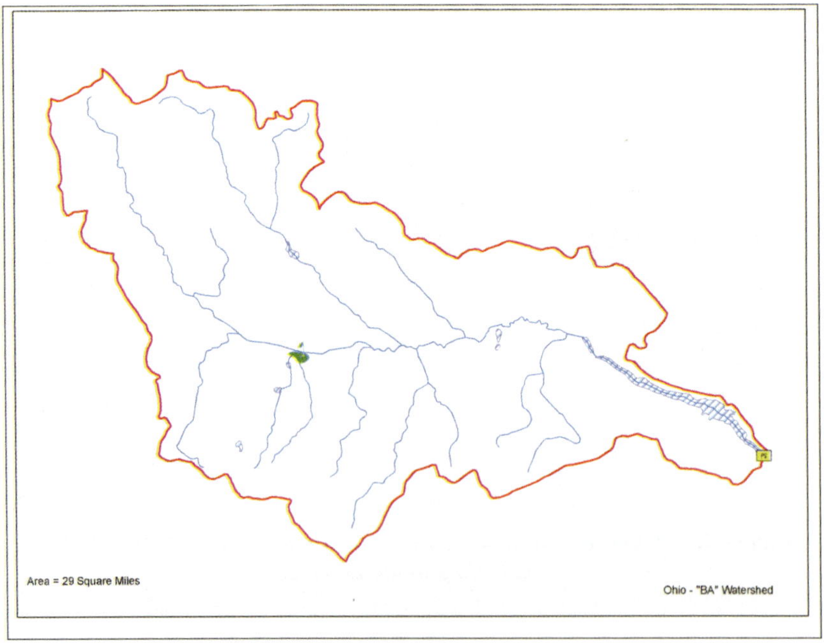

Figure 6.4. GIS presentation of the Ohio BA-VN watershed.

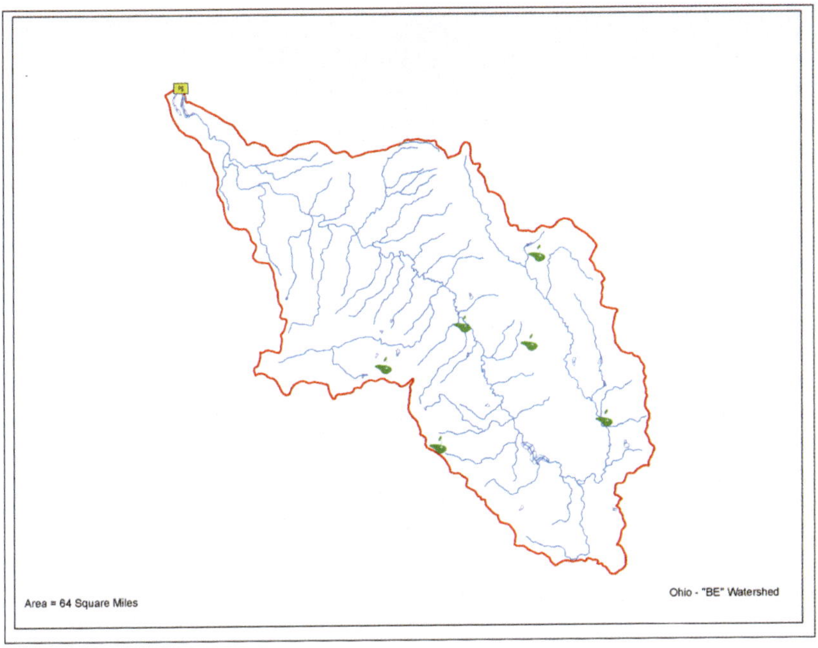

Figure 6.5. GIS presentation of the Ohio BE-VN watershed.

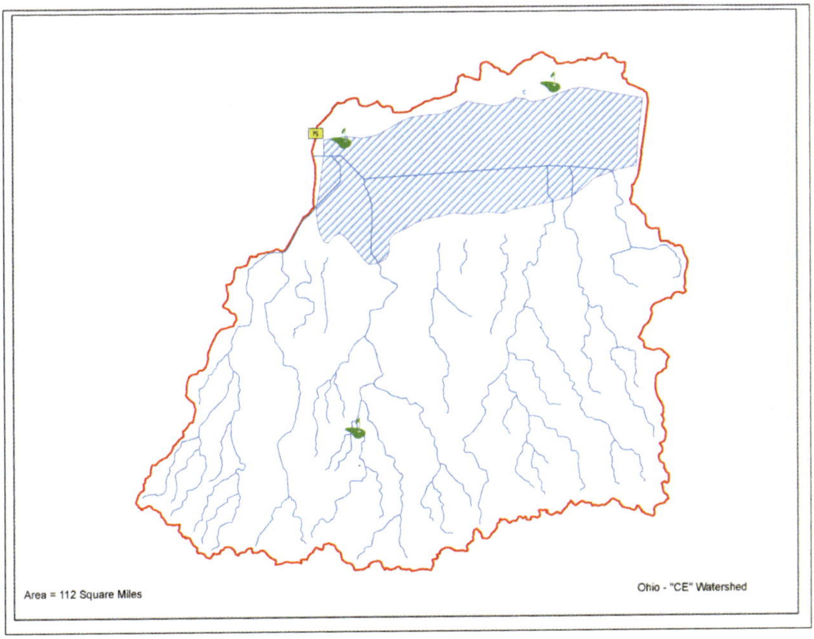

Figure 6.6. GIS presentation of the Ohio CE-VN watershed.

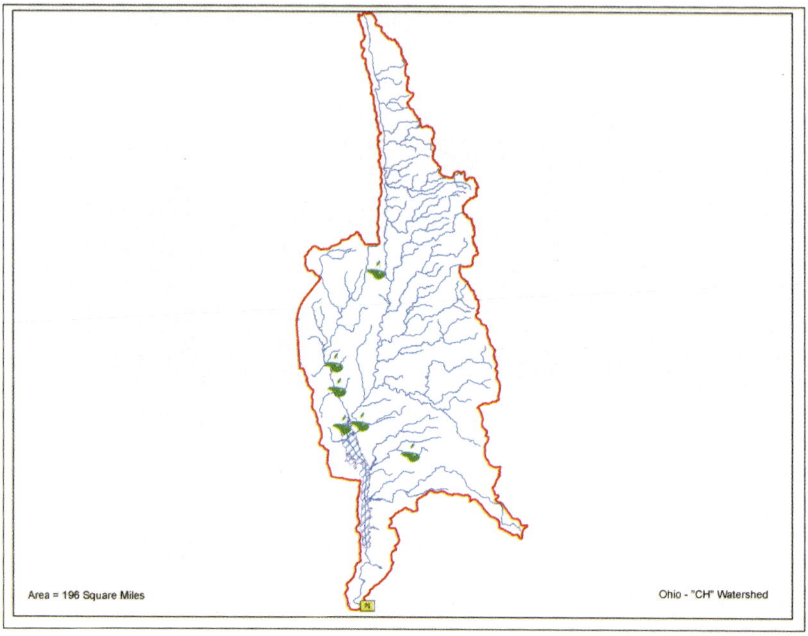

Figure 6.7. GIS presentation of the Ohio CH-VN watershed.

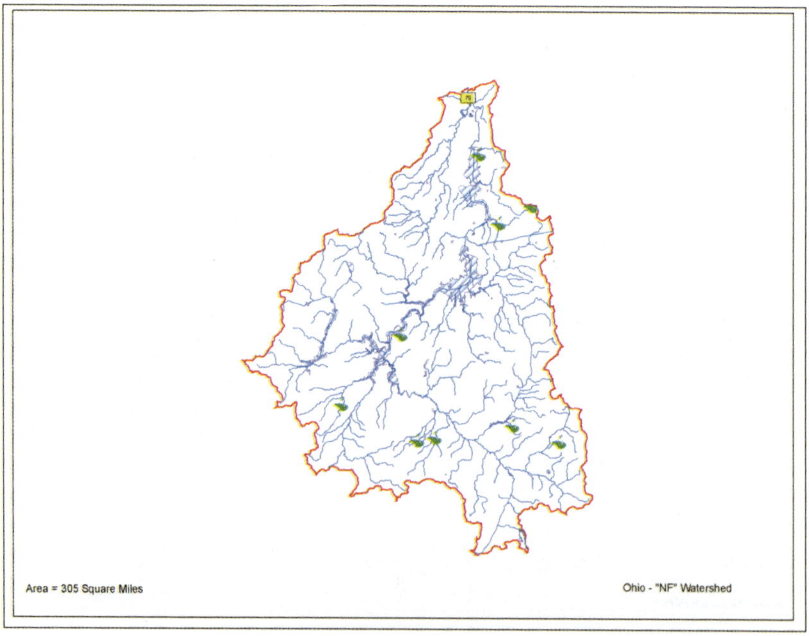

Figure 6.8. GIS presentation of the Ohio NF-VN watershed.

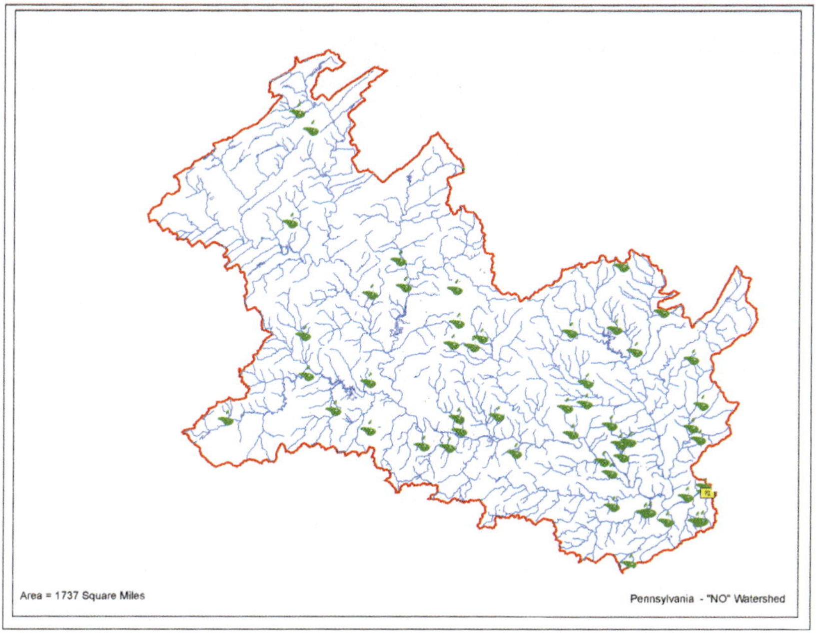

Figure 6.9. GIS presentation of the Pennsylvania EL-VN watershed.

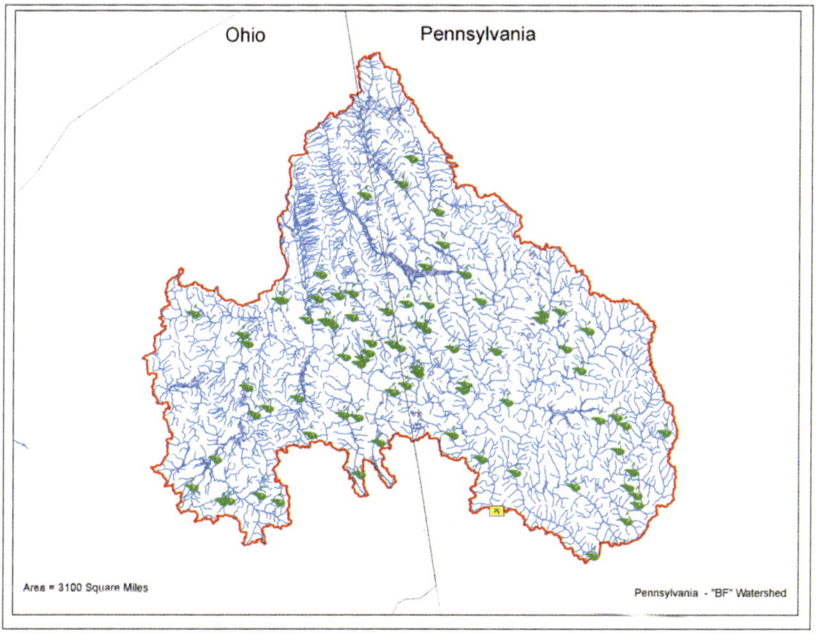

Figure 6.10. GIS presentation of the Pennsylvania NO-VN watershed.

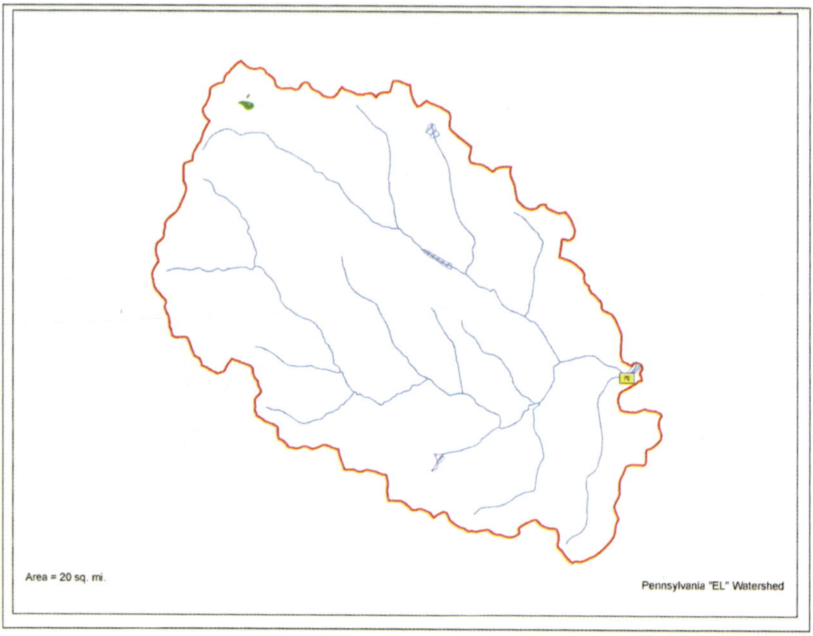

Figure 6.11. GIS presentation of the Pennsylvania PBA-VN watershed.

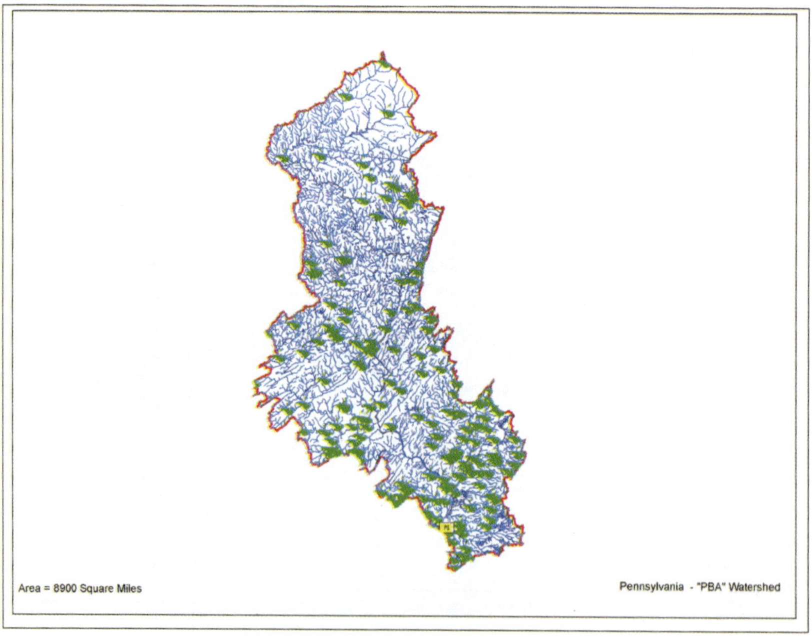

Figure 6.12. GIS presentation of the Pennsylvania PBE-VN watershed.

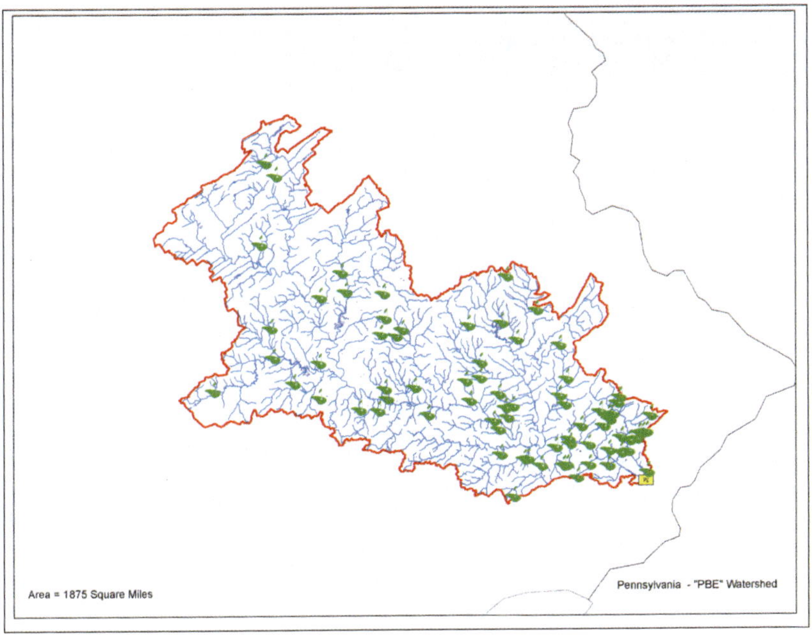

Figure 6.13. GIS presentation of the Pennsylvania BF-VN watershed.

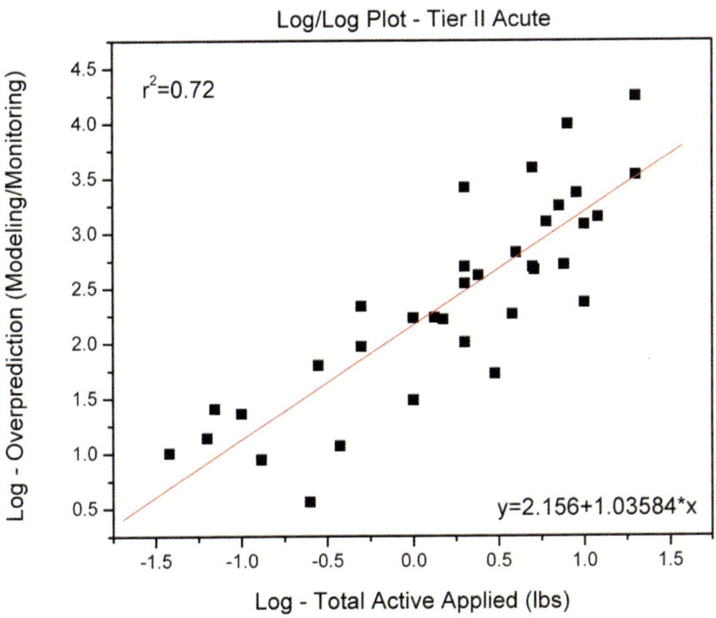

Figure 6.14. Log/log plot of the relationship between PRZM/EXAMS exposure estimates and model overprediction.

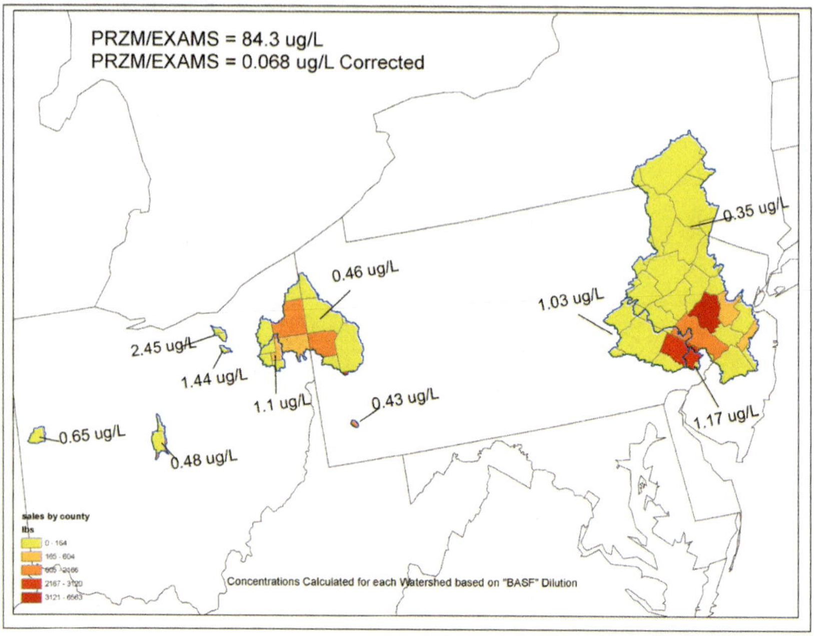

Figure 6.15. Predicted exposure concentrations in the study watersheds using the BASF dilution calculation.

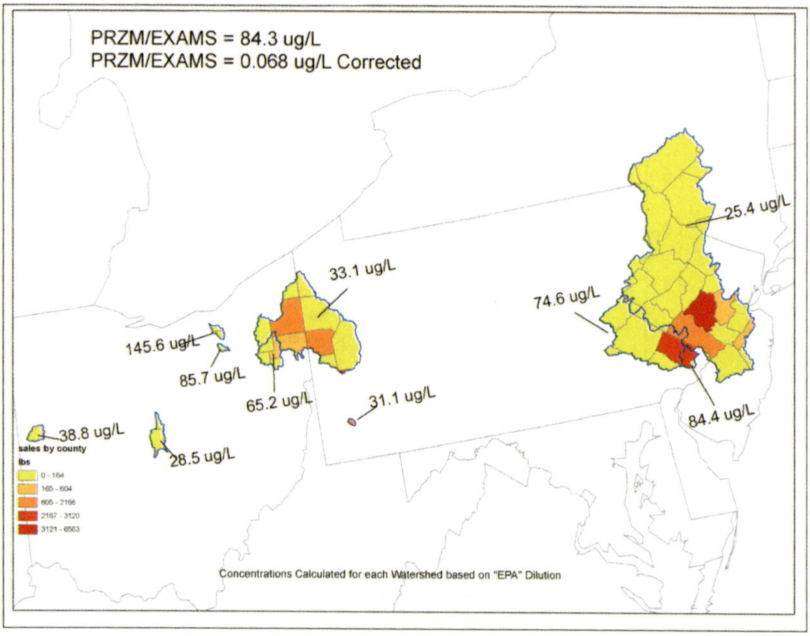

Figure 6.16. Predicted exposure concentrations in the study watersheds using the USEPA dilution calculation.

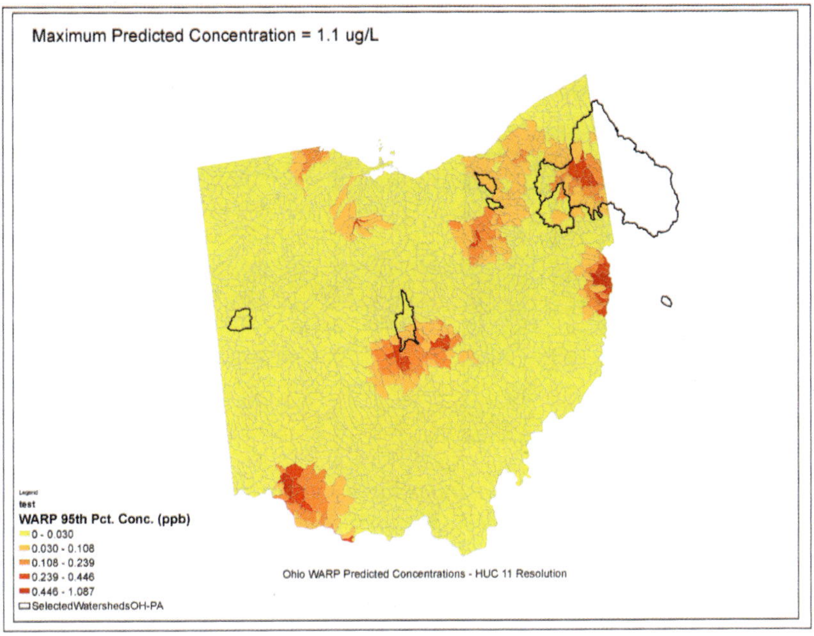

Figure 6.17. Predicted exposure concentrations in the study watersheds using WARP for the state of Ohio.

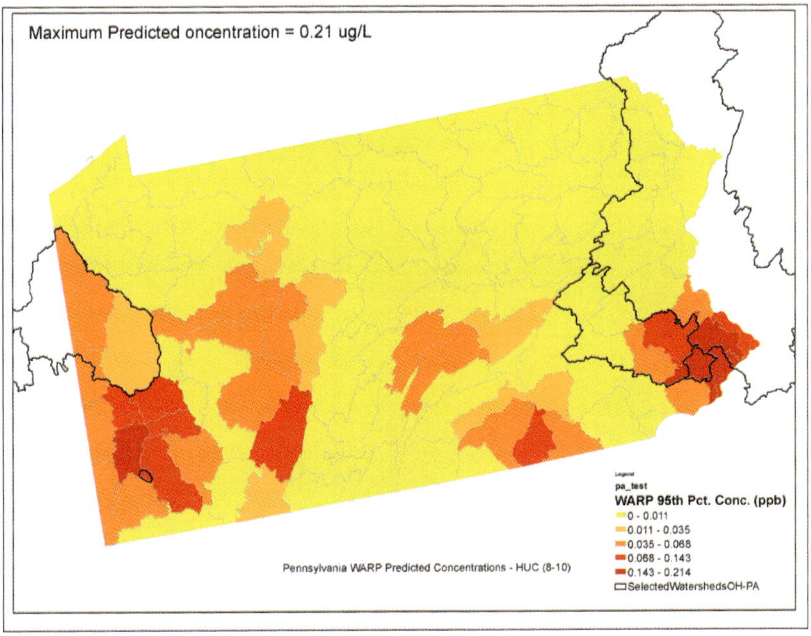

Figure 6.18. Predicted exposure concentrations in the study watersheds using WARP for the state of Pennsylvania.

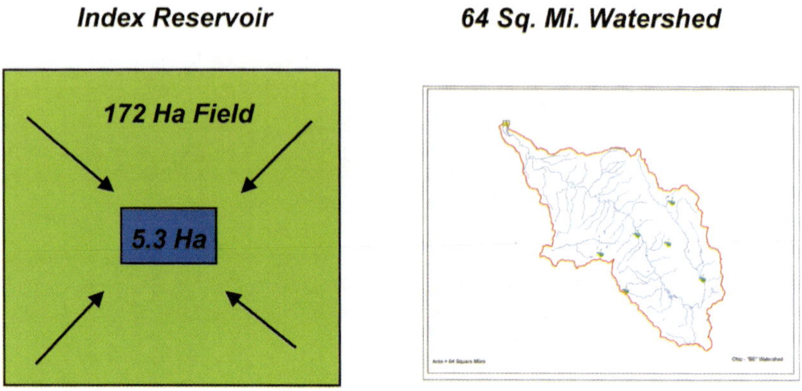

Figure 6.20. Comparison of the Index Reservoir conceptual model and a GIS coverage of an actual watershed. The arrows indicate aerial drift enters the water body from all directions.

borders should not be allowed to run onto the treated plot, just as the runoff collection apparatus must capture all surface runoff and not leak. If water external to the treated plot is allowed to run onto the plot, chemical concentrations in runoff will be diluted below their actual values. Runoff that completely bypasses or leaks from the runoff collection apparatus before measurement will reduce the total runoff volume and chemical load(s) measured. Neither of these scenarios will accurately reflect the maxium concentration and/or total load of chemical transported in the runoff that occurred.

To prevent extraneous water from entering the plot, the plot must be isolated from the surrounding areas using metal dykes (*7*), landscape timbers (*1, 8*), or flexible plastic discharge hoses filled with masonary sand (*9*). However when installing multiple, permanent plots, turf-covered soil berms may be the simplest to maintain as they can be easily mowed when less than 5-cm in height.

Plot spacing is also important and dependent on the overall experimental design and configuration of spray equipment and rainfall simulator (if any) to be used. Wide plot spacing prevents overspray during pesticide application and rainfall simulation, and allows movement of equipment between multiple plots. Knowledge of the distance of throw of the rainfall simulator is also needed to determine appropriate plot spacing.

One of the most important aspects of plot construction and maintenance is the interface that exists between the down-slope edge of the plot and the runoff collection apparatus. This interface between the runoff diverter and turf is critical because it represents a potential point of loss for surface runoff. Wauchope et al. (*5*) note that construction of the diverter-turf interface requires creativity and skill. Several approaches may be used, but in each case the system must ensure against runoff bypass and potential leaks. In Mississippi, the transition between the sod and diverter was minimized by keeping the diverter thin. Our diverter, a piece of 20-gauge aluminum metal bent at a 135° angle, was designed so that it extended into the plot by ~5-cm and extended into the runoff collection trough by ~8-cm. The soil underneath the diverter was sieved, tamped, and carefully leveled so that no air pockets were present under the diverter. The diverter was next attached to a wooden box lining the collection trench using silicone sealant and screws with rubber grommets. Prior to installing the diverter, sod close to the interface was removed using a sod cutter. Once the diverter was installed, the original sod was overlapped onto the diverter by about 3-cm. The diverter-sod interface was allowed to heal for six to eight weeks before leak testing the remaining portion of the runoff collection system using a water-soluble dye.

Maintainence of this interface, at least for bermudagrass and zoysiagrass, consisted of frequent visual inspection and clipping with hand shears as needed to keep the sod edge in good form. Line trimmers are not recommend as they can easily damage the grass-runoff divertor interface.

Weather Data Requirements

Weather data necessary for modeling purposes are generally those used to estimate evapotranspiration (ETo), namely solar radiation, air temperature, wind speed, relative humidity, and precipitation. These data must be representative of conditions at the runoff site at the time of study conduct. Thus, the weather station should be sited at, or in close proximity to, the runoff site. The guidance provided by the U.S. EPA (*10*) on siting weather stations should be closely followed to ensure the data are properly collected and suitable for modeling purposes.

Soil and Thatch Data Requirements

Input parameters needed to model turf chemical runoff vary with the intended use and capability of the model. Models that examine only runoff and rely on the curve number method to estimate this typically have few input requirements. The TurfPQ model for example, requires only inputs of chemical sorption, chemical decay and organic carbon content once a curve number has been selected for a site (*11*). Process-based models that simulate both leaching and runoff have more intensive data requirements. These models require characterization of the bulk density and moisture retention properties of thatch and soil to adequately simulate water movement in mature turf. They also require delineation of the horizons present in the soil profile, and input of the surface infiltration properties of the soil (*12, 13*).

Characterization of thatch requires that care be used in identifying and separating this medium from the underlying soil. Thatch collected for the determination of pesticide sorption coefficients and organic carbon content is best obtained using a Soil Profile Sampler or equivalent (Turf Tech International©, Tallahassee, FL). The relatively wide flat faced sample obtained with this device is superior to a cylindrical core sample for identifying and separating thatch from the underlying soil. A sharp knife can to be used to quickly separate thatch from soil when using this device.

Extracting intact (ie., undisturbed) thatch cores from the field for the purpose of determining the physical properties of this medium is more difficult than extracting soil cores. Difficulties encountered with extracting intact thatch cores from the field include the relatively unpredictable thinness or thickness of thatch found in most turf situations, the massive presence of roots at the thatch-soil interface and the highly elastic nature of the medium itself.

These difficulties can be overcome by stacking a series of thin (ie., 0.5 cm) machined half circle rings above and below an uncut ring that is placed within the barrel of a core sampler. Usually, an uncut ring having a thickness 1 to 2 cm is sufficient for determining the physical properties of thatch at most turf sites. When a core sample is extracted from the field, the half circle rings are disassembled until the thatch soil interface is clearly visible. Correct placement of the uncut ring within the barrel of the sampler is obtained by observing the location of the thatch soil interface after performing a number of core extraction trials. Ultimately, correct ring placement is dependent on how deep a core is driven into the ground.

When a core is driven to the correct depth, the thatch-soil interface is observed just below the base of the solid ring. After the half rings have been disassembled, a sharp hacksaw blade is used to section the thatch from the underlying soil. Thatch or foliage residing above the top of the ring is removed using scissors. Use of a knife or dull blade to cut through the mass of roots present at the thatch soil interface causes the thatch to pull away from the sides of the core, compromising the physical integrity of the sample. Conversely, use of a fine bladed sharp hacksaw blade facilitates a clean cut through the mass of roots and results in a core that is flush with the bottom of the ring.

Relatively few measurements of the bulk density and moisture retention properties of thatch appear in the published literature (*14, 15*). Our experience with obtaining these measurements suggest that more thatch cores need to be collected from a site than soil cores, to achieve comparable coefficients of variation for the two media (Table I). The greater coefficient of variation associated with thatch measurements is due to the lower bulk density and field capacity soil moisture retention levels found in thatch compared with soil, and because thatch core depths are generally more shallow than soil cores depths.

Table I. Physical Properties of Thatch (1 cm) and Soil Surface Cores (0 to 6 cm) Collected from Three Similarly Managed Creeping Bentgrass Plots

Media	*Bulk Density*[b] *(g/cm*^{3}*)*	*Moisture Retention*[a] *(cm*^{3}*/cm*^{3} *x 100)* 33.3 kPa	1,500 kPa	Available Water
Thatch (N = 30)	0.103 ± 0.02	24.3 ± 2.35	16.0 ± 1.89	8.30 ± 1.30
Soil (N = 24)	1.59 ± 0.06	31.5 ± 1.38	11.6 ± 0.84[c]	19.9

[a] Intact thatch cores (1cm x 5.4 cm dia.) were used to measure thatch moisture retention at 33.3 and 1500kPa, while intact soil cores (6 cm x 5.4 cm dia.) were used to determine soil surace moisture at 33.3 kPa. Disturbed sample material (N= 12) was used to determine the surface soil moisture retention at 1,500 kPa.

[b] Arithmetic mean ± one standard deviation.

[c] N = 12

Verification of Rainfall Application Rate

If rainfall is to be simulated, the delivery rate and uniformity of the rainfall simulator must be verified under field conditions. Rainfall application rates significantly less or greater than the target rate and/or lacking in uniformity may cause non-representative and/or highly variable results that greatly complicate data interpretation. Performance testing of a rainfall simulator is accomplished using a formal audit procedure (*5*). Carroll (*3*) used wide mouth plastic cups spaced on ~30-cm centers. Coefficients of uniformity should be ≥ 85% to minimize experimental error. The operating pressure of the simulator should be noted during performance audits, and checked periodically during study conduct, to ensure the system is operating properly.

Plastic tarps placed over the entire plot area (Figure 1) are useful in determining *total* rainfall delivery. This approach provides a non-quantitative, visual assessment of uniformity helpful in refining simulator design and operation. Table II demonstrates that good correlation existed between rainfall estimates provided by raingauges and a whole-plot plastic tarp. Note that tall, narrow-top rain gauges may not accurately measure rainfall, owing to the steep descent of simulated raindrops. During actual runoff events, pan-type rain gauges should be used to record actual rainfall amounts and uniformity.

Figure 1. Co-verification of simulated rainfall application rate using rain gauges (metal pans) and a whole-plot plastic tarp (foreground).

Table 2. Comparison of Two Methods Used to Verify Simulated Rainfall Application Rates (target amount was 38 mm/h).

Simulation	*Average of Six Rain Gauges*	*Whole Plot Tarp*[a]
	(rainfall amount, mm)	
1	33 ± 4	38
2	40 ± 8	41
3	39 ± 7	40
4	42 ± 4	45
5	44 ± 6	45
6	43 ± 7	37
Avg. ± std.	40 ± 6	41± 3

[a]These values represent rainfall estimated using plastic tarps that diverted entire plot runoff directly into runoff collection apparatus.

Tracking Water Flow Using a Conservative Tracer

The flow of water across the turf surface is one important determinant of the timing and extent of chemical runoff. Thus, tracking and quantifying the movement of water is important when generating runoff data for modeling purposes as this (a) aids in model calibration, and (b) helps in verifying that model predictions are reasonable. A number of inorganic anion and dye-type tracers are available for use, each having advantages and disadvantages (*16*). We found that KBr applied at 15 kg KBr/ha in 384 L H_2O/ha thirty minutes before initiation of simulated rainfall (38-mm/hr) was well tolerated by *Mississippi Pride* bermudagrass and *Myers* zoysiagrass. At this application rate, the average peak Br^- concentration in runoff was 70.7 ± 20.9 parts per million (N = 18 plots). These peak concentrations were observed within 5 min of runoff initiation. A key advantage of KBr is that Br^- can be reliably and economically analyzed by ion-selective electrode using U. S. EPA Method 9211(limit of detection ~ 0.2 ppm).

Verification of Chemical Application Rates

One must know the actual amount of pesticide(s) and tracer applied to turf, rather than assuming the nominal rate was applied. This is critical to (a) accurately calculate the percentage of pesticide(s) and tracer that occur in runoff and (b) ensure that the pesticide concentrations measured in runoff reflect those that would occur with labeled applications.

In field experiments involving pesticide application, it is not uncommon for actual application rates to differ by ± 15% or more from nominal rates, even after careful calculation, calibration of spray equipment, and application by experienced personnel (*17, 18*). In an analysis of over 1,600 pesticide applications, improper boom height (60% of errors), miscalculation of application rate (26%), and variation in pass time (14%) were determined to be most responsible for inaccurate application rates (*19*).

Three main approaches may be used to verify chemical application rates (*20*). Two indirect measures are the *catch-back* method and the *pass-time* method. The *catch-back* method involves measuring the spray solution volume before and after application to determine if the desired volume of test solution was applied to the test plots. The *pass-time* method involves measuring the time that it took the applicator to pass over a test plot of known length, and the comparison of this measured value to the nominal time used in the application target calculation. Experienced applicators are able to apply within ± 2% of the targeted spray volume or pass time; making several practice runs before each pesticide application may improve overall accuracy. An advantage of the pass-time method is that it provides realtime feedback on whether or not the application was nominally on target. Field protocols written for regulatory purposes typically require that the application be within ± 5% of the target spray volume or pass time value; variances exceeding these criteria should be closely scrutinized and the cause of the misapplication determined before proceeding with additional applications.

Our preferred verification method was to directly measure deposited residues per area, using application verification (AV) monitors. These are paper discs, polyurethane foam plugs, Petri dishes, etc. placed in the test plot to collect actual spray deposition that occurs during application. The AV monitors were collected and chemically analyzed for the test chemical(s) being applied. Pre-labeled monitors were positioned before application in an arrangement spanning the length and width of each test plot, to allow a representative sample of the spray pattern (Figure 2). We used ~ one AV monitor per 9.3-m^2 of plot area. Immediately after application, the monitors were carefully collected and handled so as to not lose their contents, wrapped in aluminum foil, and immediately frozen until analyzed.

Care must be taken not to walk on or otherwise disturb treated turf surfaces after application. A 'catwalk' or other device may be helpful in preventing plot disturbance when retrieving the AV monitors. If, after analysis ,the chemical amounts are found to vary by more than 20% within an application, the source(s) of the variability should be determined and reduced to ensure uniform pesticide and/or tracer applications in future studies (*21*).

Sample Handling & Storage

The application verification monitors and runoff samples must be properly labeled, handled and stored in order to preserve their scientific integrity. Improper handling can result in unacceptable degradation losses, and compromise the integrity of the entire study. The collection of application monitors should begin immediately after application, and the samples stored frozen to stabilize residues and solidify liquid spray droplets. Provisions should be made to have ample staff to collect the application monitors, recognizing that labor requirements rise with plot size and number of monitors used. A 'dry run' in collecting the AV monitors helps in assessing the time needed to collect, wrap and store the monitors. Sealed AV samples should be placed on wet or blue ice during or immediately after collection, and transported on ice back to the laboratory.

A robust, sensitive analytical method(s) should be in place before initiating the field-conduct phase of a runoff study, as this helps to ensure the timely analysis of samples. If all of the samples cannot be analyzed soon after collection, it may be best to analyze at least a subset of runoff samples that contain measurable residues of the targeted analytes. These samples would then be frozen along with the remaining unanalyzed samples and reanalyzed when the remainder of sample sets are analyzed. By comparing the initial and final analyses of these samples, the storage stability of residues in the later-analyzed samples can be demonstrated.

Figure 2. Co-verification of pesticide application rates using AV monitoring (Petri dishes, denoted by arrows) and pass-time methods (inset).

Summary

Much effort and expense are associated with the conduct of a turfgrass runoff experiment. While all aspects of the study are important, several are of critical importance to overall outcome of the study. Careful pre-study planning culminating in a detailed study protocol can pay tremendous dividends on studies this complex. Provisions must be made to carefully collect soil-thatch and weather data needed for modeling purposes. Use of a conservative tracer allows for improved runoff model calibration and verification of model predictions. Protection of sample integrity through all phases of study conduct is critical to the scientific validity of study results. Ultimately, an approach that tries to (a) account for as much as possible of the applied rainwater and chemical(s), and (b) minimize within-plot experimental variability through careful plot construction and maintenance, chemical application and sample handling often leads to the best overall outcomes for the researcher.

Acknowledgements

We gratefully acknowledge the financial assistance provided by the U.S. Golf Association, the Mississippi Water Resources Research Institute and the Mississippi Agricultural and Forestry Experiment Station. We also thank Dr. Scott Jackson, BASF, Research Triangle Park, NC, for his helpful comments.

References

1. Smith, A.E., and D.C Bridges. Movement of Certain Herbicides Following Application to Simulated Golf Course Greens and Fairways. *Crop Sci.* **1996**, 36: 1439-1445.
2. Lee, Young-Duek, Hyun-Ju Kim, and Jong-Bae Chung. Loss of Pendimethalin in Runoff and Leaching from Turfgrass Land Under Simulated Rainfall. *J. Agric. Food Chem.* **2000**, 48:5376-5382.
3. Carroll, M. Turfgrass Runoff Investigations: Does Plot Size Matter? *USGA Turfgrass and Environmental Research Online*. 2007, 6(24):1-7.
4. Massey, J. Bridging the Gap: Effects of Plot Size and Warm-Season Grass Species on Turf Chemical Runoff. Annual Report to the United States Golf Assn., 2007, Stilwater, OK .
5. Wauchope, R.D., R.L. Graney, S. Cryer, C. Eadsforth, A.W. Klein, and K.D. Racke. Pesticide Runoff: Methods and Interpretation of Field Studies. *Pure & Appl. Chem.* **1995**, 67: 2089-2108.
6. Nett, M.T. and Hendley P. Designing Effective Runoff Research Studies: A Review of Issues of Scale, in *Environmental Fate: Bridging the Gap Between Laboratory and Field Studies*, Phelps W. et al (eds.). ACS Symposium Series No. 813. American Chemical Society: Washington DC, 2002; pp. 23-41.
7. Wauchope, R.D., R.G. Williams, and L.R. Marti. Runoff of Sulfometuron-methyl and Cyanazine From Small Plots: Effects of Formulation. *J. Environ. Qual.* **1990**, 19:119-125.
8. Hong, S. and A.E. Smith. Potential Movement of Dithiopyr Following Application to Golf Courses. *J. Environ. Qual.* **1997**, 26:379-386.
9. Cole, J.T., J.H. Baird, N.T. Basta, R.L. Huhnke, D.E. Storm, G.V. Johnson, M. E. Payton, M.D. Smolen, D.L. Martin, and J.C. Cole. Influence of Buffers on Pesticide and Nutrient Runoff from Bermudagrass Turf. *J. Environ. Qual.* **1997**, 26:1589-1598.
10. U.S. Environmental Protection Agency. Meterological Monitoring Guidance for Regulatory Modeling Applications. Office of Air Quality, Planning and Standards, Research Triangle Park, NC 27711. EPA-454/R-99-005, 2000. (171 pgs.).
11. Haith, D.A.vTurfPQ: A Pesticide Runoff Model for Turf. *J. Environ. Qual.* **2001**, 30:1033-1039.
12. Ma, Q.L., A.E. Smith, J.E. Hook, R.E. Smith, and D.C. Bridges. Water Runoff and Pesticide Transport from a Golf Course Fairway: Observations vs. OPUS Model Simulations. *J. Environ. Qual.* **1999**, 28:1463-1473.

13. Malone, R.W., L.R. Ahuja, L. Ma, R.D. Wauchope, Q. Ma and K.W. Rojas. Application of the Root Zone Water Quality Model (RZWQM) to Pesticide Fate and Transport: An Overview. *Pest Manag. Sci.* **2004,** 60:205-221.
14. Hurto, K.A., A.J. Turgeon and L.A. Spomer. Physical Characteristics of Thatch as a Turfgrass Growing Medium. *Agron. J.* **1980**, 72:165-167.
15. Linde, D.T.,T.L. Watschke, A.R. Jarrett and J.A. Borger. Surface Runoff Assessment from Creeping Bentgrass and Perennial Ryegrass Turf. *Agron. J.* **1995**, 87:176-182.
16. Flury, M. and N. N. Wai. Dyes as Tracers for Vadose Zone Hydrology. *Rev. Geophysics*. **2003**, 41(1): 2-1 – 2-37.
17. American Crop Protection Association (APCA) Workshop Minutes,"Zero-Time Residue Levels in Field Soil Dissipation Studies." 1996, Crystal City, VA.
18. Mojasevic, M., and C.S. Helling. 9th Int. Cong. Pestic. Chem., IUPAC, Abstr. No. 2:7C-049. 1998.
19. Braverman, M.P., J.S. Corley, D.C. Thompson, M. Arsenovic, V.R. Starner, K.W. Dorscher, K.S. Samoil, F.P. Salzman, D.L. Kunkel, and J.J. Baron. Calculation and Survey of Application Accuracy. *Proc. Weed Sci. Soc. Amer*. **2001**, 41, 17.
20. Massey, J.H., S.H. Jackson, M. Saha and E. Zietz. Soil Sampling and Analysis. In P. Lee (ed.) *Handbook of Residue Analytical Methods for Agrochemicals*. John Wiley & Sons Ltd., 2002, pp. 840-891.
21. Massey, J.H. and J.S. LeNoir. Sources and magnitudes of variability in terrestrial field dissipation of pesticides. In *Terrestrial Field Dissipation Studies: Design, Interpretation and Purpose*. E. Author et al. (eds.) ACS Symposium Series No. 842. American Chemical Society, Washington, DC, 2003, pp. 73-87.

Chapter 9

Modeling Approach for Regulatory Assessment of Turf and Golf Course Pesticide Runoff

James N. Carleton[1], Cheryl Sutton[2], James Lin[2] and Mark Corbin[2]

[1]Environmental Protection Agency, Office of Water, Office of Science and Technology
[2]Environmental Protection Agency, Office of Pesticide Programs

The U.S. Environmental Protection Agency (EPA) uses the models PRZM (Pesticide Root Zone Model) and EXAMS (Exposure Analysis Modeling System) to estimate environmental concentrations (EEC) of pesticides for aquatic exposure assessments and the estimated drinking water concentrations (EDWC) of pesticides for human exposure assessments. To address turf chemicals, EPA has developed specific scenarios for PRZM-EXAMS that estimate runoff of pesticides from managed turf grass. This paper discusses these turf scenarios and associated adjustment factors that are applied to model outputs specifically for golf courses.

Objectives

The Environmental Protection Agency, Office of Pesticide Program's (EPA/OPP) "turf work group" was formed in 2000, for the purpose of developing a methodology for simulating runoff loadings of pesticides from turf-covered landscapes. The methodology would have the primary purpose of providing scientists in OPP with tools for assessing exposures to aquatic organisms, and to people who consume water impacted by runoff of turf-applied chemicals. For practical reasons, it was desirable that the methodology be implementable within the same essential modeling framework (i.e., using the current versions of PRZM and EXAMS) that is in routine use by OPP scientists to estimate pesticide runoff loadings from cropped agricultural land, and resulting concentrations in receiving waters.

Background

Using NASA satellite data, researchers have estimated the area covered by turf grass in the continental United States to be about 63,000 square miles, or 1.9% of total area (*1*), which makes turf the largest irrigated "crop" in the continental U.S., with about three times the acreage of irrigated corn. America's devotion to turf-covered landscapes exacts a high ecological toll in terms of lost habitat, and, perhaps more insidiously, in squandered opportunities to choose alternative vegetation to support struggling native fauna (e.g. bird) populations with food and shelter (*2, 3*). The energy and chemical inputs necessary to maintain turf grass covered landscapes may also induce environmental damage via nutrient and pesticide runoff (*4, 5*), and, perhaps, through emissions of the potent greenhouse gas nitrous oxide (N_2O) (*6*).

OPP is charged with assessing the health and ecological risks directly associated with the use of pesticides. For this purpose, OPP uses the Pesticide Root Zone Model (PRZM) to simulate pesticide runoff and leaching. PRZM simulates two zones in an agricultural field — the cropped zone and the soil zone. The cropped zone includes the region above the land surface. The soil zone includes the region below the land surface. Turf, unlike most agricultural crops, can have a third important zone: the thatch zone. The thatch zone is located between the cropped zone (foliage) and the soil. Thatch is made up of live as well as dead (undecomposed or partially decomposed) grass leaf and root material. Thatch is important in turf modeling because it possesses hydrologic and chemical properties which may differ significantly from the other two zones described above. The thatch zone may strongly influence movement of both water and pesticide from the surface into the soil. Correctly representing the properties of the thatch zone is therefore important to simulation of pesticide runoff and leaching from turf areas.

Scenario Development Approach

In order to meet the short-term objectives described above, the team revisited an unpublished modeling approach developed by James Lin at Bayer in the early 1990's which employed the existing version of the model (PRZM 1). This involved adding thatch as a 2 cm. layer of "soil" on top of an actual soil profile, similar to an approach later used by Duborow et al. (*7*) to model pesticide runoff from turf. The following critical thatch properties needed to model it as soil were obtained from an associated laboratory study (unpublished) on Kentucky bluegrass thatch: field capacity = 0.47; wilting point = 0.27; organic carbon = 35.6%. A value for bulk density of 0.37 g/cm^3 was obtained from a published study (*8*). Results from a small turf plot runoff study were used to back-calculate a curve number for the site, and PRZM was run to simulate the runoff of pesticides under the artificial rainfall conditions of the study. Differences between modeled total pesticide runoff loads and measured loads leaving the four plots ranged between -27.4% and 34.5%, indicating fair agreement between model predictions and data.

A key consideration in modeling thatch as a soil layer is the selection of an appropriate value for % organic carbon (%OC). Although thatch has a very high organic carbon content, pesticide sorption to organic carbon in thatch is not well characterized by the results of sorption studies conducted in soils. Several studies that have examined and compared K_{oc} values for pesticides in thatch and in soil have found lower K_{oc} values in thatch (*9, 10*). This may be due to the relatively undecomposed nature of the organic matter in thatch, and resultant differences in hydrophobicity of carbon in thatch as compared to soil. Unfortunately, OPP does not typically have studies of pesticide sorption on thatch to develop model inputs and must use sorption coefficients derived from studies on soil.

The approach adopted by the OPP Turf Work Group expanded on Dr. Lin's basic approach, and in the absence of thatch-based K_{oc} values, used soil-based K_{oc} values (which are available in registrant-submitted environmental fate studies) to model thatch sorption. In this approach, PRZM was first calibrated via curve number adjustment so that modeled runoff matched the experimentally observed runoff response to a set of artificial rainfall events. Empirically-based adjustments to the %OC in modeled thatch were then made so that the sorptive behavior of studied pesticides (as reflected in the total mass of pesticide lost from the field in overland runoff) reasonably matched the results obtained in published small plot turf runoff studies.

Small Plot Studies Used to Calibrate Effective %OC in Thatch

The results of published studies conducted in the Piedmont Region of Georgia (*11, 12*) were found to contain sufficient detail to calculate a value for effective %OC in the thatch layer. These studies involved small-plot simulated golf course fairways (planted in bermuda grass) to which 2,4-D, dicamba, mecoprop, and dithiopyr were applied. The soil at the site was described as a Cecil sandy clay loam. Simulated rainfall of known volume (2.5 to 5 cm) was applied on days 1, 2, 4 and 8 after pesticide application. Total water volume leaving the plots after each artificial rainfall was reported in one study (*12*). Total pesticide load leaving the plots in runoff was reported in both studies.

In order to simulate these applications and runoff events in a manner consistent with standard EPA approaches, the meteorological file that would ordinarily be used for modeling this specific region [Major Land Resource Area (MRLA) 136] was altered to include these "rainfall" additions on the specified dates. Extra rainfall was added to the file in arbitrarily-selected years 1956 and 1957, since the meteorological file did not include data for the time period after 1983, and the actual applications took place in 1993 and 1994. The soil profile in the PRZM input file was developed as described above, with soil layer thicknesses and properties for a Cecil sandy clay loam obtained from the Data Base Analyzer and Parameter Estimator (DBAPE) database (*13*), and a generic 2-cm thick layer of "thatch" on top as described above. Application rates and dates were set to match those of the actual applications. The PRZM foliar extraction coefficient (0.5) and pesticide fate properties were set in accordance with existing OPP input parameter guidance. Maximum rooting depth was set at 3-cm, so that roots extended 1 cm below the thatch layer into the soil. The model was run for 1956, using various trial values of nominal curve number (CN2) until the total volume of runoff for the simulated events matched the observed volume reasonably well (Figure 1), which occurred with CN2 set to 93. This is is similar to the value of 91 that Durborrow et al. (*7*) found fit data from this site.

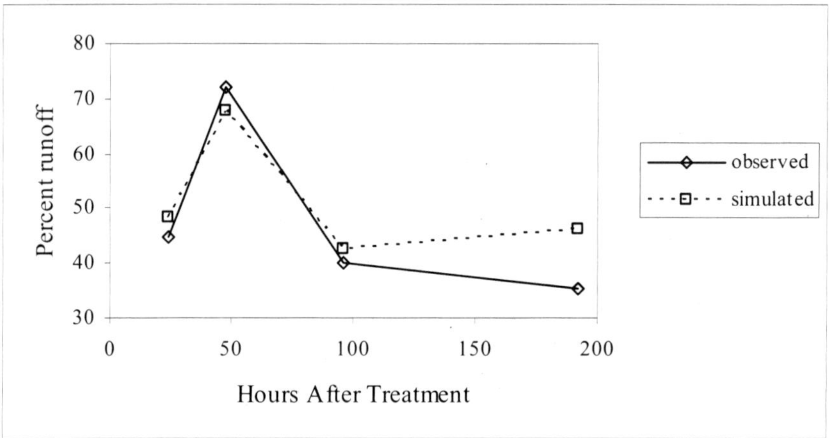

Figure 1. Modeled vs. observed runoff at Georgia small plot turf runoff study, shown as a function of hour after treatment.

With the curve number set at 93, PRZM was then run for 1956 and 1957, using various trial values of %OC in the thatch layer, to model two compounds with high and low mobilities: 2,4-D (K_{oc}=34.23) and dithiopyr (K_{oc}=1920), respectively. By trial and error, a value of 7.5% OC was found to result in good agreement between model predictions and the runoff data for both compounds (Figures 2, 3, 4). Results for compounds modeled using K_d rather than K_{oc} (dicamba, K_d=0.07 mL/g; and mecoprop, K_d=0.29 mL/g) also matched the GA data reasonably closely (Figures 5, 6). Note that because K_d was used instead of K_{oc} to model these latter two compounds, sorption was modeled as independent of %OC, and the same results would have been obtained with any assumed %OC value for the "thatch" in the PRZM input file. This simply indicates that, for low-sorbing compounds, the chemical runoff algorithms already present in PRZM provide adequate representations of runoff in these particular small plot studies, without the need for pseudo-empirical adjustment of soil parameters to account for the presence of thatch.

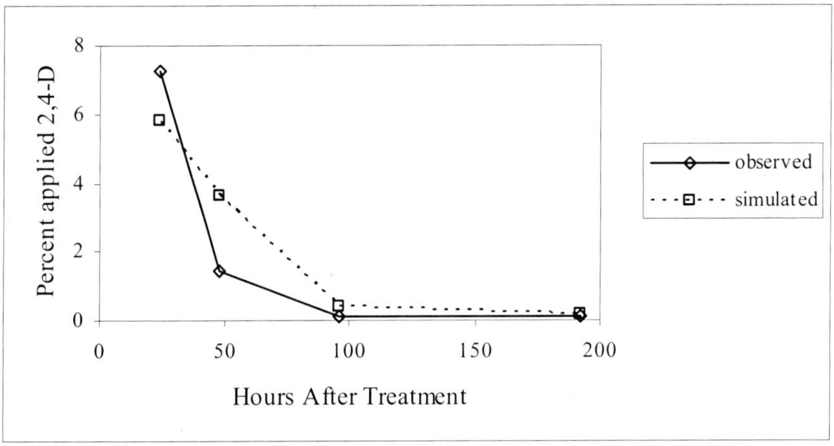

Figure 2. Modeled vs. observed 2,4-D runoff loading at Georgia small plot turf runoff study, shown as a function of hour after treatment.

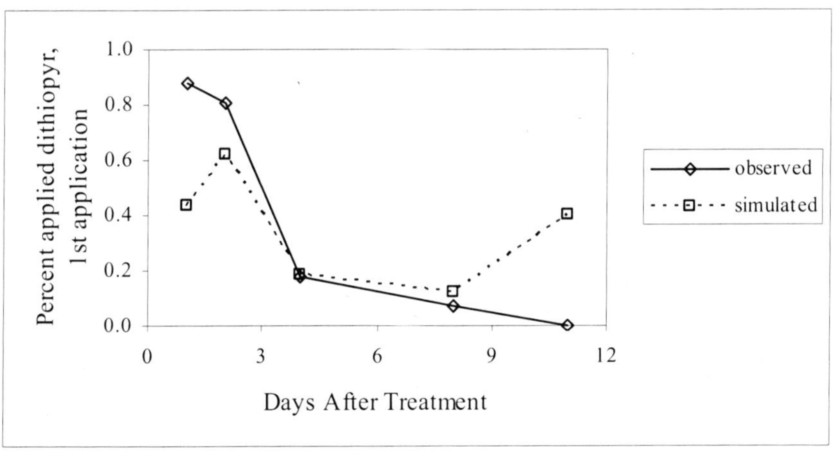

Figure 3. Modeled vs. observed dithiopyr runoff loading at Georgia small plot turf runoff study (first application), shown as a function of days after treatment.

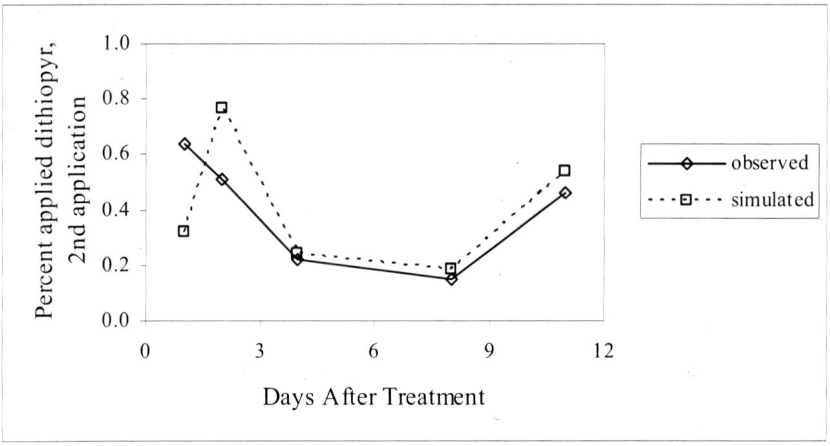

Figure 4. Modeled vs. observed dithiopyr runoff loading at Georgia small plot turf runoff study (second application), shown as a function of days after treatment.

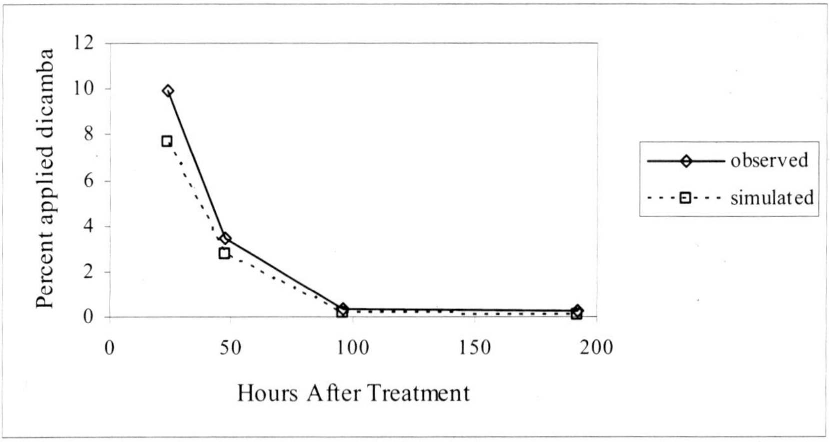

Figure 5. Modeled vs. observed dicamba runoff loading at Georgia small plot turf runoff study, shown as a function of hours after treatment.

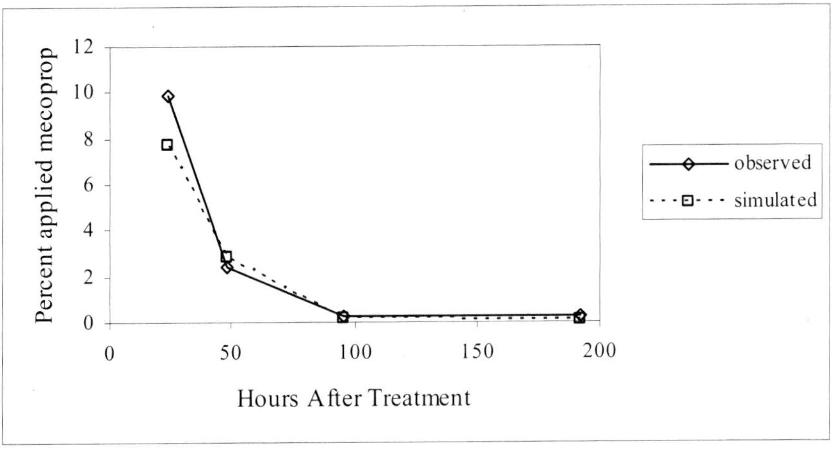

Figure 6. Modeled vs. observed mecoprop runoff loading at Georgia small plot turf runoff study, shown as a function of hours after treatment.

Summary of Approach

The approach developed by OPP for constructing turf runoff modeling scenarios was to select soils (and their properties) for the region to be modeled, just as one would do to develop an agricultural runoff scenario. A 2-cm deep layer of "thatch" was then added on top of the modeled soil profile, posessing the following properties: bulk density = 0.37; field capacity = 0.47; wilting point = 0.27; organic carbon = 7.5%. Curve numbers were selected based on "good condition" open space areas as specified in TR-55 (*8*), that is, 39, 61, 74, and 80 for hydrologic soil groups A, B, C, and D, respectively. A 2-cm layer of thatch is typical for golf course fairways, but is probably thicker than average for golf course greens (Mike Kenna, personal communication; Research Director, United States Golf Association Green Section). Modern greens built according to current USGA specifications are designed to rapidly infiltrate water, and are built upon sand/peat mixtures, with tile underdrainage. However, a large fraction of the greens in the United States are of the old-style "push-up" variety, composed essentially of existing soil from the site, and lacking underdrainage. In the interests of simplicity, transparency, and implementability, turf was considered to be essentially generic, with no distinction made between fairways, greens, tees, or residential lawns. For chemicals applied to golf courses, the fraction of the total area composed of greens, tees, and fairways may be used to modify the results of a modeling run, somewhat in the fashion of a percent cropped area (PCA) adjustment for agricultural chemicals.

Adjustments for Golf Course Turf

OPP uses a tiered system for drinking water exposure assessments. At Tier I, screening-level models are used to assess pesticide concentrations in drinking water. Tier I is designed to screen out chemicals with low potential risk for posing a drinking water concern. If the Tier I exposure estimates are determined to represent unacceptable exposure, then a more refined Tier II assessment is conducted which provides more site-specific, refined estimates by taking into account additional environmental fate parameters, specific soil data, weather information, and management practices to estimate daily concentrations of pesticides in water for an extended period of time (up to 30 years). In both Tier I and Tier II assessments, surface water results are adjusted for a PCA factor which takes into account the fact that only a portion of the watershed being modeled may be planted to the specific crop being modeled. However, PCAs are only applicable to agricultural crops, not to non-food use crops such as turf. In cases where a pesticide is used only on golf course turf, additional adjustment factors were needed to account for the percent acreage of a golf course (and, thus, a watershed made up entirely of golf course land) that is not treated with a pesticide. The Golf Course Adjustment Factor (GCAF) was used to refine surface water EDWCs and EECs generated by OPP's aquatic exposure models for golf course turf scenarios.

Golf course facilities consist of separate playing areas that are classified as tees, greens, practice greens, fairways, driving ranges, and roughs, in addition to "unmanaged grounds" where lakes, ponds, out-of-play areas, conservation areas, and buildings are located. Depending on the playing area, management practices and intensity can vary in these facilities. When an individual pesticide is used, for example, only on tees and greens, or tees and greens plus fairways, if it is assumed in the modeling scenario that the entire golf course is treated, this assumption can lead to overestimation of the EDWCs and EECs. The use of the GCAF to refine those values can correct for this by quantitatively discounting the percentage of managed land area on a golf course that is not treated with a pesticide.

Background Information

Based on the World Golf Foundation's "The Golf 20/20 Industry Report for 2002," (*15*), there were about 15,827 golf facilities in the United States as of March 2003. An average-sized 18-hole golf facility is about 150 acres of total land (including water bodies, hard structures and out-of-play areas), of which *ca.* two-thirds are maintained turf (*16*). A typical urban golf course is only 110-120 acres, and courses in resort areas may be 170-190 acres (Greg Lyman, personal communication 11/19/04; Director of Environmental Programs, Golf Course Superintendents Association of America). Generally, pesticides are not applied to entire golf courses, but rather to some holes and some parts of the course (e.g., tees, greens, and/or fairways). They may be applied as spot treatments or to an entire portion of a course, although pesticide labels are rarely specific on the usage details. Tees and greens are typically the most intensively managed areas, and tend to receive higher pesticide inputs compared with fairways and roughs.

In determining EDWCs for a drinking water assessment, OPP utilizes a standard EXAMS scenario referred to as the "index reservoir" in Tier II modeling with PRZM/EXAMS. OPP's Tier I model FIRST (FQPA Index Reservoir Screening Tool) also uses the index reservoir scenario, which simulates a 172.8-ha field (watershed) draining into a 5.3-ha reservoir. For agricultural crops, a PCA is used to adjust the EDWCs to account for the portion of a drinking water watershed containing fields planted with a specific crop. PCAs have been developed for only a few agricultural crops to date, largely limited by availability of crop acreage data at the required scale. At the present time, data are not available at the scale needed to finalize development of the non-agricultural equivalent of a PCA for golf course uses.

For agricultural and non-agricultural crops, it is assumed that the entire field is treated. While this is often the case for agricultural crops, it is not typically the case for golf courses. Thus, for a drinking water exposure assessment, the GCAF is used to adjust the EDWCs to account for the percentage of the field that is not treated. This, in effect, makes the GCAF a "percentage land area treated" adjustment, characteristic of land use on a golf course. This adjustment is applied to both Tier I and Tier II modeling outputs for use in a drinking water assessments.

In determining EECs for an aquatic ecological exposure assessment, OPP uses a standard EXAMS scenario referred to as the "small static pond" in Tier II modeling with PRZM/EXAMS. Scenarios simulate a 10-hectare field draining into a 1-hectare static pond that is 2-meters deep and does not have an outlet. The pond serves as a surrogate for the range of small, sensitive water bodies that can be found in the headwaters of a watershed, including low-order streams. It is assumed that runoff is equally likely to flow into the pond from all areas of the treated field, and that the entire field is treated. With the small pond and turf scenarios, OPP concluded that EECs were representative of a subset of ponds that occur on golf courses, given their configuration, the size of the ponds and their drainage areas. Thus, a GCAF is not used with Tier I EECs from GENEEC2 (Generic Estimated Environmental Concentration Model, v. 2.0), the Tier I model used to determine pesticide concentrations in surface water for aquatic ecological exposure assessments. It is only used after Tier II (unadjusted) EECs result in relevant Level of Concern (LOC) exceedences. If there are no exceedences, then only the risk quotients (RQs) derived from non-modified EECs are reported in the risk assessment. For ecological exposure modeling, a GCAF is only used to refine the Tier II EECs. In this case, both the adjusted and unadjusted Tier II EECs are reported in the ecological risk assessment. This approach differs from that used in estimating drinking water exposure because, for drinking water, effects are integrated over a watershed of larger spatial scale and estimated concentrations are generally accepted as reasonably conservative.

Supporting Data for Recommended GCAF Values

In developing this GCAF guidance, two independent sources of data were reviewed. Additional information/data searching was conducted by personal communication and by consulting research reported on golf organization Web sites (e.g., Golf Course Superintendents Association of America at www.GCSAA.org and the United States Golf Association at www.USGA.org. The data obtained from the GCSAA, presented in Table 1, were utilized to develop the recommended GCAFs. The data from a USGA survey (which was an internal survey for use by pesticide registrants) were not used to develop the GCAFs, as the results were based on information for a smaller sample of golf courses and percentages were calculated based on total facility acreage, including non-turfgrass areas.

The survey data provided by the GCSAA was based on the survey responses from 741 GCSAA members submitted over two years. Responses represented multiple course types including private, semi-private, daily fee, municipal, resort, and other. The majority of the courses were 18-hole facilities (572); others that were represented included 9-hole (79) and ≥19-hole (90) facilities. Respondents were from eight USGA regions of the country: Northeast (90), Mid-Atlantic (58), Southeast (73), Florida (45), Mid-Continent (150), North Central (163), Northwest (36) and Southwest (122). The distribution of responses matched the distribution of GCSAA members by course type, size and USGA region. Survey data used to develop the GCAFs are presented in Table I.

Table I. Golf Course Superintendent Association of America Golf Course Acreage Survey Data

Use Type	*Average Number of Acres*[a]	*Percentage of Course*[b] *(%)*
Tees	2.7	2.4
Greens	2.9	2.6
Fairways	31.9	28.6
Roughs	66.8	60.0
Practice Green[c]	0.2	0.018
Driving Range[d]	7.1	6.4

[a]Based on personal communication with Greg Lyman, GCSAA, 11/19/2004, these data represent the final results of the GCSAA 1999-2000 survey for golf course size.

[b]Percentage represents course subtype divided by total maintained turf acreage of 111.5 acres. Acreage in lakes, ponds, out-of-play areas and hard structure acreage is not included; when included, the average size of a golf course is closer to 150 acres.

[c]Practice green acreage is managed similar to greens and is accounted for in the recommended GCAF for tees and greens.

[d]Driving range acreage is managed similar to roughs and is accounted for in the recommended GCAF for roughs.

OPP Procedure for Use of Adjustment Factors Specific to Golf Course Turf on Tees, Greens, Fairways and Roughs

This procedure describes how OPP adjusts EDWCs (drinking water assessment; both Tier I and Tier II) and refines EECs (ecological exposure assessments; Tier II only) resulting from pesticide use on golf course turf. This modification of estimated concentrations, using OPP aquatic exposure models, accounts for the percentage of the total golf course acreage that actually receives pesticide treatment.

For pesticides applied only to tees and greens, the FIRST or PRZM/EXAMS output values are multiplied by 0.05 to modify the EDWCs (surface water only) and Tier II EECs, and the resulting values are reported as the adjusted EDWCs or EECs. For applications to fairways only, the output values are multiplied by 0.29. When tees, greens, and fairways are all treated, the output values are multiplied by 0.34. If tees, greens, fairways and roughs are all treated, a GCAF is not utilized, as the output values are "multiplied by" a factor of 1. For EECs, adjusted and unadjusted concentrations are both reported. The GCAFs are summarized in Table II.

Table II. Recommended Golf Course Adjustment Factors by Turf Type

Treated Areas of Course (Turf Type)	*GCAF*
Tees & Greens (includes practice green)	(0.024 + 0.026) = 0.05
Fairways	0.29
Roughs (includes driving range)	0.66
Tees & Greens & Fairways	(0.05 + 0.29) = 0.34
Tees & Greens & Fairways & Roughs	(0.05 + 0.29 + 0.66) = 1.0

Use/Guidance Restrictions

The assumption that the entire watershed of a drinking water reservoir is comprised of golf course land, and that all of this land is treated, is a conservative simplification that is necessary given land use and pesticide application data limitations. With approximately 15,827 golf facilities in the United States, the co-occurrence of golf courses and other crops treated with the same pesticide in the same watershed cannot be discounted. OPP's default PCA of 0.87 for agricultural crops cannot be used to refine EDWCs for golf course or other turf uses because the number was derived without including turf acreage. The current default PCA is based on the highest percentage of agricultural land in any United States Geological Survey (USGS) 8-digit Hydrologic Unit Code (HUC) in the conterminous United States. More data at a relevant spatial scale are necessary before OPP can develop PCAs for turf grass or other non-agricultural uses, including golf courses.

The GCAF is different from the PCA factor, which is applicable only to agricultural crops and refers to the fraction of a watershed that is planted with a particular crop. Instead, it is a correction factor used to account for the partial land area treated relative to the total golf course acreage. The GCAF is applied to model estimates for drinking water assessments. For ecological exposure modeling, the GCAF is only used to refine the Tier II EECs if Tier II (unadjusted) EECs result in exceedences of relevant LOCs. The GCAF is not used for Tier I EECs from GENEEC2. If there are no exceedences in Tier II modeling, then only the RQs derived from non-modified EECs are reported in the risk assessment. If there are exceedences in Tier II modeling, both the adjusted and unadjusted Tier II EECs are reported in the ecological risk assessment, and the differences are described in the risk characterization. The GCAF is not applied to groundwater values determined using the Tier 1 model SCI-GROW (Screening Concentration in Ground Water), which is based on a treated field rather than a watershed.

The adjustment factor is only applicable to golf course use scenarios, and is not used for other turf use scenarios, such as sod farm, residential, right-of-ways, (other) recreational or any other turf uses. If other uses are permitted on the label(s), the adjustment factor is used only to modify, as appropriate, the values reported for golf course turf use. If there are any turf uses on the label other than golf course turf (for example, if the use is for "undifferentiated turf" or "sod farms"), the unadjusted values are reported to represent those uses, in addition to the adjusted values representing golf course use. Additionally, when used for modifying EECs, both the initial Tier II EEC and adjusted/refined Tier II EEC values are reported in the risk assessment.

Remaining Uncertainties

While a GCAF allows the user to modify the EDWCs and EECs determined by aquatic exposure models for a golf course turf use scenario, it does not take into account all uncertainties involved in estimating surface water concentrations associated with the use of a pesticide on a golf course.

There are several aspects of pesticide use on golf course turf that may result in the model scenario underestimating surface water concentrations. Golf courses are commonly built near water; many are near wetlands. Golf courses are typically designed to drain water, incorporating a mix of areas with higher slopes, depressions, and tile drainage systems. These drains rapidly transport water that infiltrates to discharge points in nearby surface water bodies. Currently, neither the Tier I nor the Tier II aquatic exposure models used by OPP can account for subsurface drainage on golf courses. The turf standard scenario used in the Tier II model is a general one that is not specific to golf courses.

The GCAF takes into account the fact that not all of the turf is treated; thus, it only allows for a "percentage land area treated" adjustment. The GCAF does not account for the fact that, for drinking water, it is highly unlikely that an entire drinking water watershed would be comprised of golf course turf. Additional data, analogous to the data used to develop PCA factors, are needed to address this issue.

The values utilized to develop the GCAF represent average values, as data were not available to represent higher percentile values. Also, the use of the GCAF assumes that the estimated surface water concentrations will be reduced in equal proportion to the reduced level of acreage treated; supporting data for this assumption are not available.

Future research needs to help address some of these uncertainties include review of reliable pesticide monitoring data from golf courses, with adequate ancillary data to allow comparison with initial and refined EECs. The results of this research and additional data may lead to revised scenarios in the future.

Acknowledgements

We are grateful to John Ravenscroft, Dana Spatz, Ron Parker, R. David Jones and Elizabeth Behl for their contributions.

Disclaimer

Any opinion, findings, conclusions or recommendations expressed in this manuscript are those of the authors alone, and do not necessarily reflect the views of U.S. Environmental Protection Agency or the U.S. Government. Mention of trade names or commercial products does not constitute endorsement or recommendation for use.

References

1. Milesi, C. , Running, S.W., Elvidge, C.D., Dietz, J.B., Tuttle, B.T., and Nemani, R.R. Mapping and Modeling the Biogeochemical Cycling of Turf Grasses in the United States. *Environmental Management.* **2005**, 36(3):426-438.
2. Chapman, K.A., and Reich, P.B. Land Use and Habitat Gradients Determine Bird Community Diversity and Abundance in Suburban, Rural and Reserve Landscapes of Minnestoa, USA. *Biological Conservation.* **2007**, 135: 527-541.
3. Tallamy, D.W. *Bringing Nature Home: How Native Plants Sustain Wildlife in Our Gardens*. 2007, Timber Press, Portland, OR.
4. Waschbusch, R.J., Selbig, W.R., and Bannerman, R.T. *Sources of Phosphorus in Stormwater and Street Dirt From Two Urban Residential Basins in Madison, Wisconsin, 1994-95.* 1999, USGS Water-Resources Investigations Report 99-4021.
5. Weston, D.P., Holmes, R.W., and Lydy, M.J. (2008) Residential Runoff as a Source of Pyrethroid Pesticides to Urban Creeks. *Environmental Pollution*. 2008, in press.
6. Bremer, D.J. Nitrous Oxide Fluxes in Turfgrass: Effects of Nitrogen Fertilization Rates and Types. *Journal of Environmental Quality.* **2006**, 35:1678-1685.

7. Duborow, T.E., Barnes, N.L., Cohen, S.Z. Horst, G.L. and Smith, A.E. Calibration and Validation of Runoff and Leaching Models for Turf Pesticides, and Comparison with Monitoring Results, in *Fate and Management of Turfgrass Chemicals*, ACS Series 743, J.M. Clark and M.P. Kenna, eds., American Chemical Society, Washington, D.C., 2000, pp. 195-227.
8. Hurto, K.A., Turgeon, A.J., and Spomer, L.A. Physical Characteristics of Thatch as a Turfgrass Growing Medium. *Agronomy Journal*. **1980**, 72:165-167.
9. Dell, C.J., Throsell, C.S., Bischoff, M., and Turco, R.F. Organic Chemicals in the Environment: Estimation of Sorption Coefficients for Fungicides in Soil and Turfgrass Thatch. *Journal of Environmental Quality*. **1994**, 23:92-96.
10. Lickfeldt, D.W. and Branham, B.E. Organic Chemicals in the Environment: Sorption of Non-ionic Organic Compounds by Kentucky Bluegrass Leaves and Thatch. *Journal of Environmental Quality*. **1995**, 24:980-985.
11. Hong, S. and Smith, A.E. Potential Movement of Dithiopyr Following Application to Golf Courses. *Journal of Environmental Quality*. **1997**, 26:379-386.
12. Smith, A.E. and Bridges, D.C. Movement of Certain Herbicides Following Application to Simulated Golf Course Greens and Fairways. *Crop Science*. **1996**, 36(6):1439-1445.
13. Imhoff, J.C., Carsel, R.F., Kittle, J.L, and Hummel, P. R. *Database Analyzer and Parameter Estimator (DBAPE) Interactive Computer Program User's Manual*. 1990, EPA/600/3-89/083.
14. USDA Soil Conservation Service. *Urban Hydrology for Small Watersheds. Technical Release 55*, 2nd edition, June, 1986.
15. GCSAA. *Performance Measurement Survey, 1999-2000 - Preliminary Results*. February 2001.
16. Lyman, G.T., Throssell, C.S., Johnson, M.E., Stacey, G.A., and Brown, C.D. Golf Course Profile Describes Turfgrass, Landscape and Environmental Stewardship Features. *Applied Turfgrass Science*. 2000, Doi:10.1094/ATS-2007-1107-01-RS.

Chapter 10

The Development of a Standard Turf Scenario:

Notes of an External Review of the USEPA Turf Scenario

Adrian M. Wadley[1], Mary T. Nett[2,*]

[1]San Francisco CA 94122
[2]Water Quality Consulting, Colorado Springs CO 80906

The use of agricultural field scale runoff models to predict pesticide runoff losses from turf poses some unique modeling challenges. There are a number of aspects of turf grass culture which differ significantly from row crop agriculture in how they affect runoff processes. In particular, the generally high tiller and leaf density of the verdure reduces exposure of the soil surface to the direct impact of rain droplets and hence, mitigates the potential for agrochemical transport in runoff. The development of increased surface soil organic matter levels, in varying stages of decomposition and incorporation into the soil matrix as 'thatch', should also be taken into account. Highly managed turf grass areas susceptible to runoff are generally smaller than the typical field size for agricultural row crops; the potential for soil erosion is therefore, again, much reduced under turf. The United States Environmental Protection Agency (USEPA) evaluated data from a small-plot turf runoff study published by researchers at the University of Georgia (UGA) to enable development of a standard turf scenario, now in regulatory use. The USEPA scenario was then evaluated by the authors using the same publications, i.e., the small-plot data generated on the transport of the herbicides mecoprop, dicamba, dithiopyr and 2,4-D in runoff, with the addition of more detailed, unpublished data from the University of Georgia study files. In addition, the UGA-measured transport of chlorpyrifos, a strongly adsorbed compound, was included in the present evaluation. Model performance using the key features of the standard turf scenario was generally good for the weakly adsorbed compounds. Performance for the more strongly adsorbed compounds was poor, and over-predicted adsorbed chemical losses in eroded soil. Adjustments to soil erosion parameters and consideration of scale effects resulted in improved predictions.

Introduction

In 2002, the Environmental Fate and Effects Division of the USEPA (USEPA-EFED) published the findings of an internal work group established to investigate methods for modeling the potential for pesticide exposure of surface waters following chemical application to turf grass (*1*). The results determined by Carleton et al were encapsulated into two turf scenarios (for Florida and Pennsylvania), and designed for use in Tier II exposure assessments with the Pesticide Root Zone Model and the Exposure Analysis Modeling System (PRZM-EXAMS) (*2, 3*). A document presented to the Exposure Modeling Work Group (EMWG) by the USEPA work group outlined the approach taken to develop the turf scenarios, and the justification for the parameterization of the exposure assessment models.

The development of the turf scenarios focused on data extracted from two publications by Professor Emeritus Albert Smith, *Movement of Certain Herbicides Following Application to Simulated Golf Course Greens and Fairways* (*4*) and *Potential Movement of Dithiopyr Following Application to Golf Courses* (*5*). The current authors, with the cooperation of the original researchers, were provided access to the comprehensive suite of data generated by these studies. The availability of previously unpublished hydrology and transport data enabled the ideas put forward by USEPA in developing the turf scenarios to be examined in more detail, using more runoff events and additional chemicals not originally reported in the literature. This extended the range of chemical properties considered from weakly sorbed herbicides to chemicals with a greater propensity to adsorb to soil and organic matter. The parameterization of the small plot turf runoff studies at the Griffin campus of the University of Georgia will herein be referred to as the Georgia Turf Scenario, to distinguish it from the actual Tier II scenarios in regulatory use for Florida and Pennsylvania

The Georgia Turf Scenario was developed using the published literature values and PRZM3 (1999) (*6*). In order to estimate the volume of water leaving a known area of land during a runoff event, PRZM implements the United States Department of Agriculture Soil Conservation Service "Curve Number" approach. The USEPA turf work group considered the hydrology of the plots, and an appropriate curve number was chosen to match the percentages of applied water as runoff reported in the UGA studies. Several pesticide applications were made to the small turf plots during 1993 and 1994, and simulated rain events of a targeted 2.5 cm and 5.0 cm were generated using sprinkler irrigation. Of the water applied, between 35.5 and 72.1% ran off. In the initial publication (*4*), these values were reported as an average of the simulated rain events across multiple test plots. In Hong and Smith (*5)*, these events were reported individually. The single curve number chosen by the USEPA work group to represent transport from all the plots and events reported in the UGA publications was 93, and this generated excellent agreement between observed and modeled runoff as a percent of nominal applied 'rainfall' (irrigation). Figure 1 illustrates the excellent agreement achieved between predicted average runoff and measured average runoff (exclusive of any chemical transport).

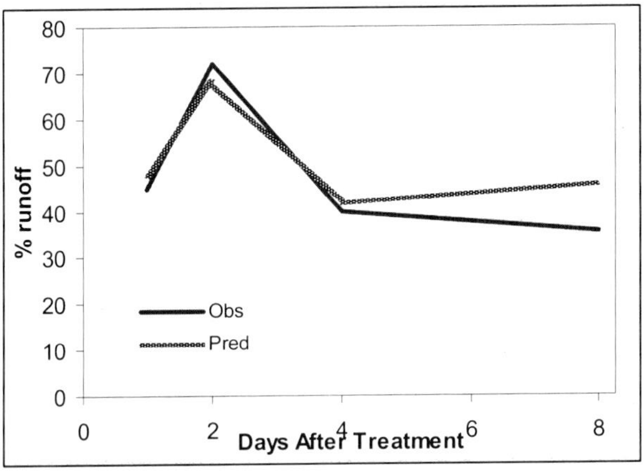

Figure 1: Observed vs predicted runoff. Observed values are averages of three replicates and three application dates. Observed data from Smith and Bridges (1996) and predicted data from Carleton, Lin and Corbin (2001).

Examination of the study files revealed some variability between plots and events, and, in one case, the actual runoff which occurred was less than that reported. These event–by–event values were useful in the re-examination reported here.

Following the choice of an appropriate curve number, the USEPA work group examined the model predictions of chemical transport in runoff. Using the approach of Hurto (*7*), they surmised that, under turf, the development of a layer enriched with organic matter, commonly known as thatch, would be effective in reducing chemical runoff. The additional organic matter derived from decaying grass clippings, stems and roots near the surface of the turf test plots provided a greater opportunity for agrochemicals, particularly for lipophilic compounds, to be retained by adsorption. However, this organic matter was relatively young and had not been fully incorporated into the soil matrix, thus the full potential for adsorption might not have been realized. The organic matter was also not fully decomposed, and had not yet had a chance to be physically broken down by the action of soil fauna. Near the surface, thatch components are more readily recognized as parts of grass plants; lower down, in the verdure/thatch/soil profile, the thatch becomes less recognizable as parts of plants, and is more incorporated into the soil. This gradation of decomposing organic matter was assumed not to provide as effective an adsorption potential as the organic matter found in well developed soils, and thus Carleton et al proposed that Hurto's figure of 36% organic carbon (OC) in thatch was too high. Through calibration of the chemical runoff values from the Georgia Turf Scenario, the effective organic carbon percentage in a topmost 2 cm deep soil horizon was established as 7.5%. Figure 2 shows the chemical runoff observations and predictions using this effective OC percentage.

The work group demonstrated good agreement between the observed and model-predicted values, and the effective organic carbon percentage of 7.5% in the 2 cm modified soil layer they proposed to simulate a thatch layer was adopted as part of the turf modeling methodology for the Tier II USEPA standard scenario. The inclusion of a modified soil layer to simulate thatch is the only part of the original modeling by the turf work group that the authors found was strongly unique in the regulatory Turf Scenario. The curve numbers, erosion parameters and the size of the treated areas reflect other aspects of the standard USEPA agricultural scenarios such as location, and typical row-crop practices and treated areas.

Revised Modeling of the Georgia Turf Scenario

The detailed UGA study notes and data were used to identify individual test chemical applications, discrete runoff events and individual turf test plots in the original research. Many of these data were presented as averages in the literature, and this averaging did not reveal plot to plot variabilities in runoff characteristics, volume, chemical transport and applied irrigation/simulated rainfall. In the original research program, a suite of twelve turf grass test plots was utilized, three of which received applications of the herbicides 2, 4-D, mecoprop and dicamba, as reported in Smith and Bridges (*4*). Discrete records of the applications to each test plot were available to the authors for individual consideration against chemical transported; in the publication utilized by the USEPA work group, these data were averaged. In addition, the other nine test plots received applications of other turf chemicals, with varied physicochemical properties. These data proved interesting for further study.

An iterative approach to attempt to replicate the original USEPA Turf Scenario work was adopted by the authors, beginning with a review of the original turf work group PRZM input files provided to us by USEPA. The assumptions inherent in parameter choices were examined, altered and re-run, but with comparison to data generated from the discrete UGA test plots, and individual events, rather than the averaged data. As far as possible, where values could be identified in the study files, actual weather and irrigation amounts were used, rather than nominal or "target" amounts.

Five turf grass agrochemicals were evaluated. Three compounds adsorbed weakly onto soil/organic matter; the remaining two more were more strongly bound, as measured by K_{oc} or K_d. Table I lists the adsorption values used by the authors in this modeling.

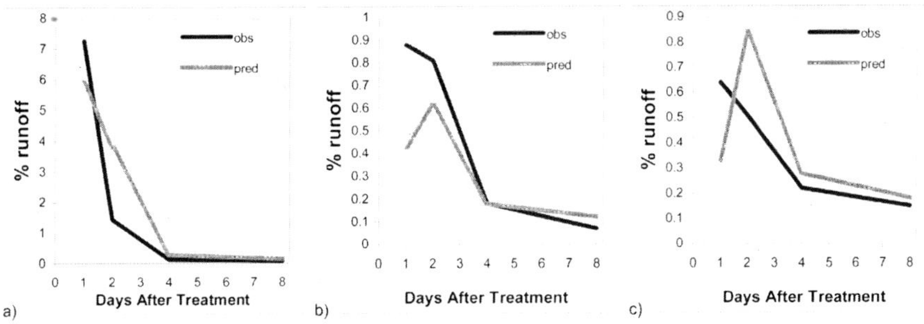

Figure 2: Predicted vs observed chemical runoff percentages for the Georgia Turf Scenario. Observed data for 2,4-D from Smith and Bridges (1996), dithiopyr data from Hong and Smith (1997) and predicted data from Carleton, Lin and Corbin (2001).

Table I. Chemicals Studied and Adsorption Coefficients

Chemical	*Adsorption Coefficient*	*Value*
Dicamba	K_d	0.07 L kg^{-1}
Mecoprop	K_d	0.29 L kg^{-1}
2, 4-D	K_{oc}	34 L kg^{-1}
Dithiopyr	K_{oc}	1920 L kg^{-1}
Chlorpyrifos	K_{oc}	9930 L kg^{-1}

Minor modifications to the modeling files obtained from the USEPA turf work group were required for the authors to achieve workable model runs. Examination of the input files revealed that there was no change in decay rate with depth, nor a different decay rate in the thatch layer. Neither of these, however, is likely to be critical for the modeling of runoff in the period immediately following application.

Of more interest was that the herbicides mecoprop and dicamba were modeled with their K_d specified, rather than their K_{oc}. Thus, the effect of the additional organic carbon in the thatch layer was ignored for these two chemicals. In the USEPA modeling, 2, 4-D and dithiopyr were modeled using their K_{oc}s, and thus the effect of the modified thatch layer was included.

The land area modeled was 10 hectares in all cases, which reflected the standard pond scenario used in USEPA ecological exposure calculations for the typical farm pond. Although the land area modeled in PRZM does not have a significant impact on chemical runoff predictions in cases where the chemicals are not transported adsorbed to eroded soil, it is significant where erosion is a concern. The comparison of predicted and observed results should be undertaken by modeling similar plot sizes when erosion may be a significant loss route for the applied chemical. The results are likely to be unaffected by this assumption for 2, 4-D, dicamba and mecoprop, which are highly soluble in water and only weakly sorbed to soil/organic matter. For these three chemicals, the majority of the chemical lost from the treatment area will be transported dissolved in the runoff stream.

The results of the PRZM modeling, using the original USEPA parameterization, were examined in three areas – water runoff, chemical transport in runoff and sediment erosion (Figures 3 through 12). As far as available, plot by plot comparisons were made. Records of the plot-specific applied irrigation (as simulated rainfall) were not available, so the UGA published averages (5 cm, 5 cm, 2.4 cm and 3.6 cm, at 24, 48, 96 and 192 hours, respectively, after treatment) were used.

Plot-specific runoff volumes were available. Significant plot to plot variability, for example, was established for a single event on July 19th, 1994 when runoff ranged from 0.4 to 1.7 cm of water for a nominal 5 cm applied. Later work by Armbrust and Peeler (*8*) showed that the irrigation equipment used to generate simulated rain storms on these plots delivered from 4.0 to 6.8 cm of water for a nominal 5.0 cm event. Turf test plots 2, 3, and 7 (the plots which received the applications of mecoprop, dicamba and 2, 4-D), on average received greater that the target amounts of simulated rainfall in subsequent studies where plot-specific water delivery was recorded. It is likely, though unknown, that a similar pattern of water delivery occurred in the published 1993 and 1994 research.

Soil erosion, as modeled by PRZM, was predicted to be high, even though these were fully grass covered plots. PRZM, as initially configured, predicted soil erosion losses in runoff water in the gram per liter range, suggesting that significant soil loss would have occurred. Soil loss predictions were in the thousands of kilograms per hectare range; in reality, measured soil erosion losses from well managed turf would be significantly less.

Water runoff, as evaluated by a comparison of the twelve plot average for each event to the modeled result, showed good agreement. However, the field data showed that the discrete plots treated with mecoprop, dicamba and 2, 4-D (Plots 2, 3, and 7), despite receiving greater than the target volume of applied rainfall, had lower than average measured runoff. Model predictions of runoff for these plots were therefore in excess of measured for the specific plots where these chemicals were applied, using the original curve number of 93. Dithiopyr and chlorpyrifos were significantly over-predicted, by up to 20 fold.

CHEM	DICAMBA		Jun-94			Jul-94			
P/O	Source/Run	Note	1daa	2daa	4daa	1daa	2daa	4daa	5daa
Observed	Field Notes	Study File	**9.1%**	**2.5%**	**0.4%**	**5.7%**	**3.5%**	**0.2%**	**0.1%**
Predicted	DICcec01	JC file	7.3%	2.9%	0.2%	7.3%	2.8%	0.2%	0.0%
Predicted	DICcec04	AW mod	7.4%	2.9%	0.2%	7.4%	2.9%	0.2%	0.0%
Predicted	DICcec10	AW mod2	8.6%	2.5%	0.1%	8.6%	2.4%	0.1%	0.0%
Predicted	DICcec14	AW3 CN86	4.3%	0.9%	0.0%	3.5%	0.9%	0.0%	0.0%

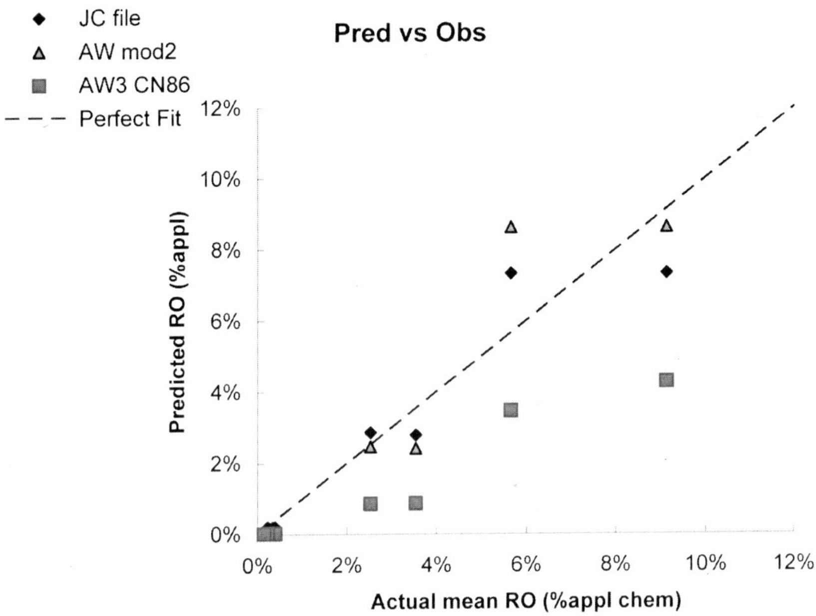

MODEL FIT		NRMSE
DICcec01	JC file	32 %
DICcec04	AW mod	32 %
DICcec10	AW mod2	40 %
DICcec14	AW3 CN86	76 %

JC file	as received
AW mod	changes to USLE only
AW mod2	soil values reflect study file
AW3 CN86	soil, met (irri+rain) from study file, CN=86 to fit hydrology

Figure 3. Dicamba – Model-predicted versus observed runoff from UGA small-plot turf study.

DICAMBA			1994	Jun-94			Jul-94			
P/O	Source/Run		Note	6/14/94	6/15/94	6/17/94	7/18/94	7/19/94	7/21/94	7/22/94
			DAA	1	2	4	1	2	4	5
Observed	Front Page	Study File	Chem RO%	9.1%	2.53%	0.40%	5.7%	3.54%	0.23%	0.15%
Predicted	DICcec01	JC file	Pred 1 (JC)	7.3%	2.86%	0.20%	7.3%	2.79%	0.20%	0.04%
Predicted	DICcec04	AW mod	Pred 2 (AW)	7.4%	2.95%	0.21%	7.4%	2.87%	0.21%	0.05%
Predicted	DICcec10	AW mod2	Pred 3 (AW)	8.6%	2.46%	0.13%	8.6%	2.41%	0.13%	0.03%
Predicted	DICcec14	CN86	Pred 5 (AW)	4.3%	0.86%	0.02%	3.5%	0.87%	0.02%	0.01%

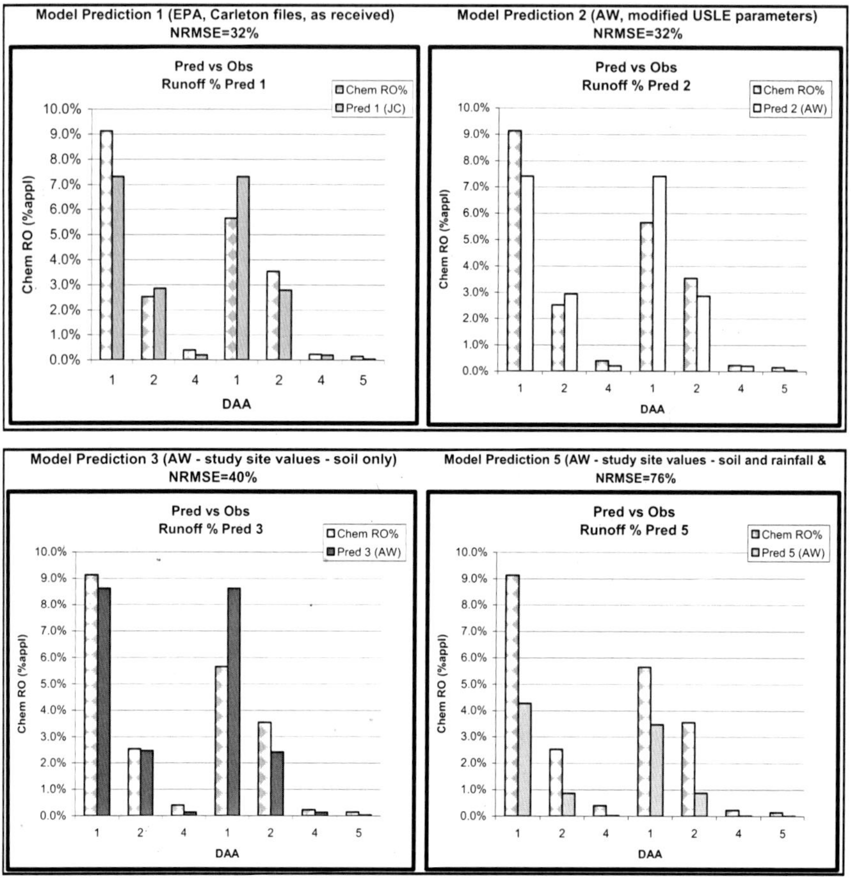

Figure 4. Dicamba – Evaluation of USEPA Turf Scenario using model predictions of transport in runoff.

CHEM	MECOPROP		Jun-94			Jul-94			
P/O	Source/Run	Note	1daa	2daa	4daa	1daa	2daa	4daa	5daa
Observed	Field Notes	Study File	**11.4%**	**3.0%**	**0.4%**	**5.4%**	**0.9%**	**0.1%**	**0.0%**
Predicted	MECcec01	JC file	7.3%	2.9%	0.2%	7.3%	2.8%	0.2%	0.0%
Predicted	MECcec04	AW mod	7.3%	3.1%	0.2%	7.3%	3.0%	0.2%	0.1%
Predicted	MECcec10	AW mod2	7.5%	3.1%	0.3%	7.5%	3.1%	0.3%	0.1%
Predicted	MECcec14	AW3 CN86	3.9%	1.1%	0.0%	3.2%	1.1%	0.0%	0.0%

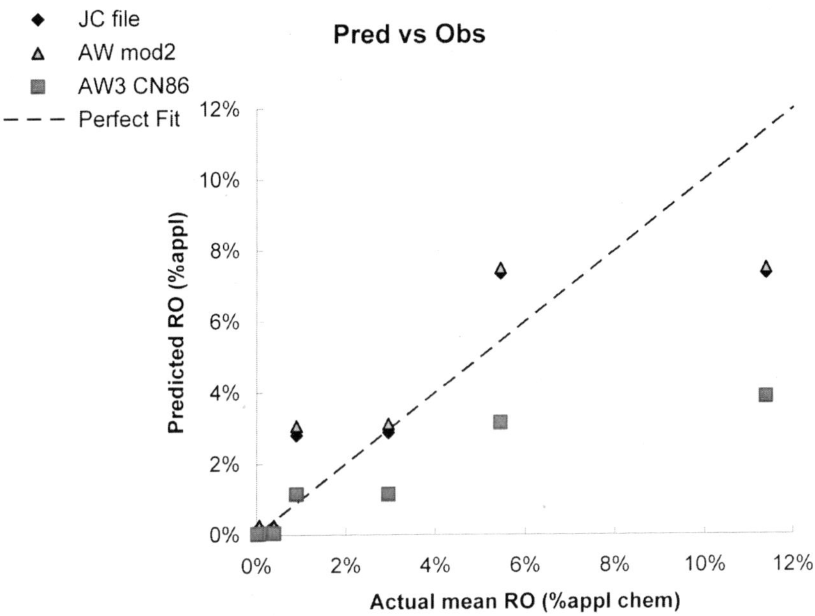

MODEL FIT		NRMSE
MECcec01	JC file	61 %
MECcec04	AW mod	63 %
MECcec10	AW mod2	62 %
MECcec14	AW3 CN86	101 %

JC file as received
AW mod changes to USLE only
AW mod2 soil values reflect study file
AW3 CN86 soil, met (irri+rain) from study file, CN=86 to fit hydrology

Figure 5. Mecoprop – Model-predicted versus observed runoff from UGA small-plot turf study.

MECOPROP			1994	Jun-94			Jul-94			
P/O	Source/Run		Note	6/14/94	6/15/94	6/17/94	7/18/94	7/19/94	7/21/94	7/22/94
			DAA	1	2	4	1	2	4	5
Observed	Front Page	Study File	Chem RO%	11.4%	3.0%	0.4%	5.4%	0.9%	0.1%	0.0%
Predicted	MECcec01	JC file	Pred 1 (JC)	7.3%	2.9%	0.2%	7.3%	2.8%	0.2%	0.0%
Predicted	MECcec04	AW mod	Pred 2 (AW)	7.3%	3.1%	0.2%	7.3%	3.0%	0.2%	0.1%
Predicted	MECcec10	AW mod2	Pred 3 (AW)	7.5%	3.1%	0.3%	7.5%	3.1%	0.3%	0.1%
Predicted	MECcec14	CN86	Pred 5 (AW)	3.9%	1.1%	0.0%	3.2%	1.1%	0.0%	0.0%

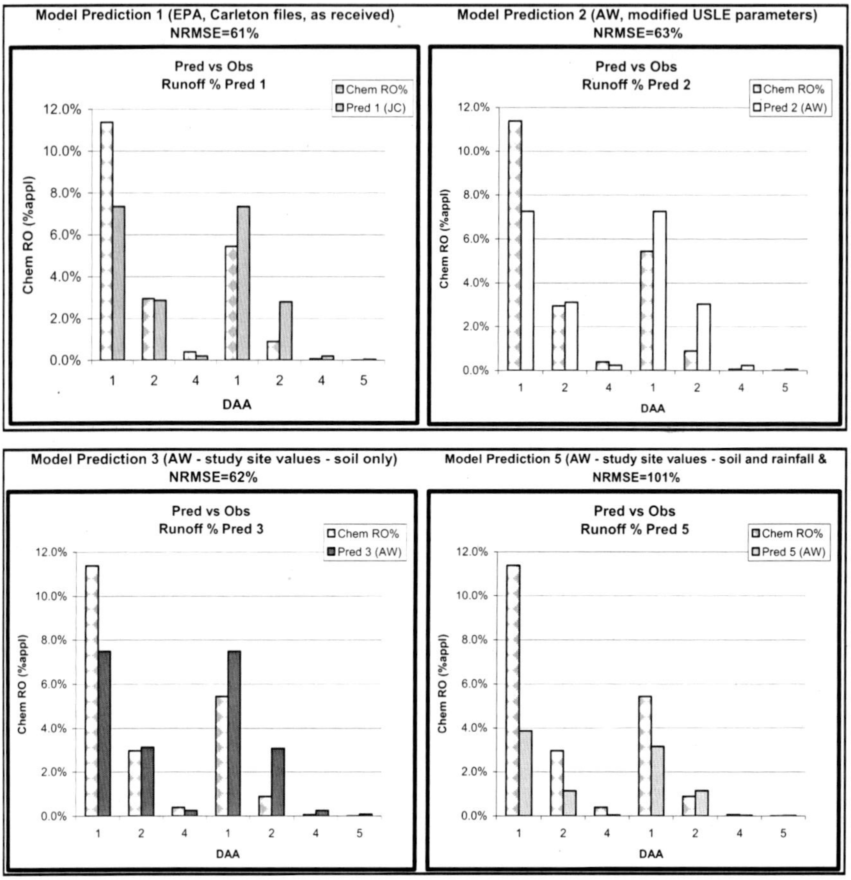

Figure 6. Mecoprop – Evaluation of USEPA Turf Scenario using model predictions of transport in runoff.

CHEM	2,4-D		Jun-94			Jul-94				Jun-94
P/O	Source/Run	Note	1daa	2daa	4daa	1daa	2daa	4daa	5daa	1daa
Observed	Field Notes	Study File	6.8%	1.5%	0.11%	4.6%	0.26%	0.03%	0.00%	
Predicted	DICcec01	JC file	5.6%	3.7%	0.42%	3.6%	0.4%	0.14%	0.00%	0.0001
Predicted	DICcec04	AW mod	5.7%	3.8%	0.43%	3.7%	0.4%	0.14%	0.00%	0.0001
Predicted	DICcec10	AW mod2	4.7%	2.3%	0.22%	4.7%	2.3%	0.22%	0.08%	0.0004
Predicted	DICcec14	AW3 CN86	2.5%	0.9%	0.04%	2.1%	0.9%	0.03%	0.03%	0.0018

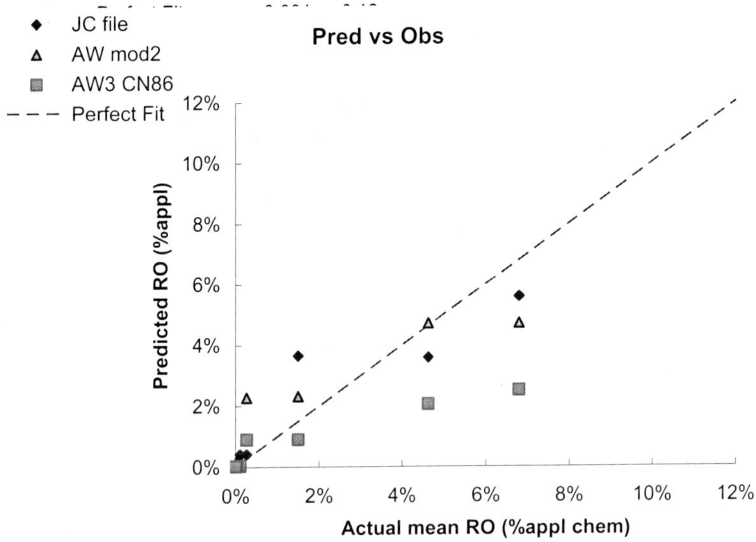

MODEL FIT		NRMSE
DICcec01	JC file	54 %
DICcec04	AW mod	54 %
DICcec10	AW mod2	60 %
DICcec14	AW3 CN86	101 %

JC file	as received
AW mod	changes to USLE only
AW mod2	soil values reflect study file
AW3 CN86	soil, met (irri+rain) from study file, CN=86 to fit hydrology

Figure 7. 2, 4-D – Model-predicted versus observed runoff from UGA small-plot turf study.

2,4-D			1994	Jun-94			Jul-94			
P/O	Source/Run		Note	6/14/94	6/15/94	6/17/94	7/18/94	7/19/94	7/21/94	7/22/94
			DAA	1	2	4	1	2	4	5
Observed	Front Page	Study File	Chem RO%	6.8%	1.50%	0.11%	4.6%	0.26%	0.03%	0.00%
Predicted	DICcec01	JC file	Pred 1 (JC)	5.6%	3.66%	0.42%	3.6%	0.42%	0.14%	0.00%
Predicted	DICcec04	AW mod	Pred 2 (AW)	5.7%	3.76%	0.43%	3.7%	0.43%	0.14%	0.00%
Predicted	DICcec10	AW mod2	Pred 3 (AW)	4.7%	2.32%	0.22%	4.7%	2.29%	0.22%	0.08%
Predicted	DICcec14	CN86	Pred 5 (AW)	2.5%	0.90%	0.04%	2.1%	0.91%	0.03%	0.03%

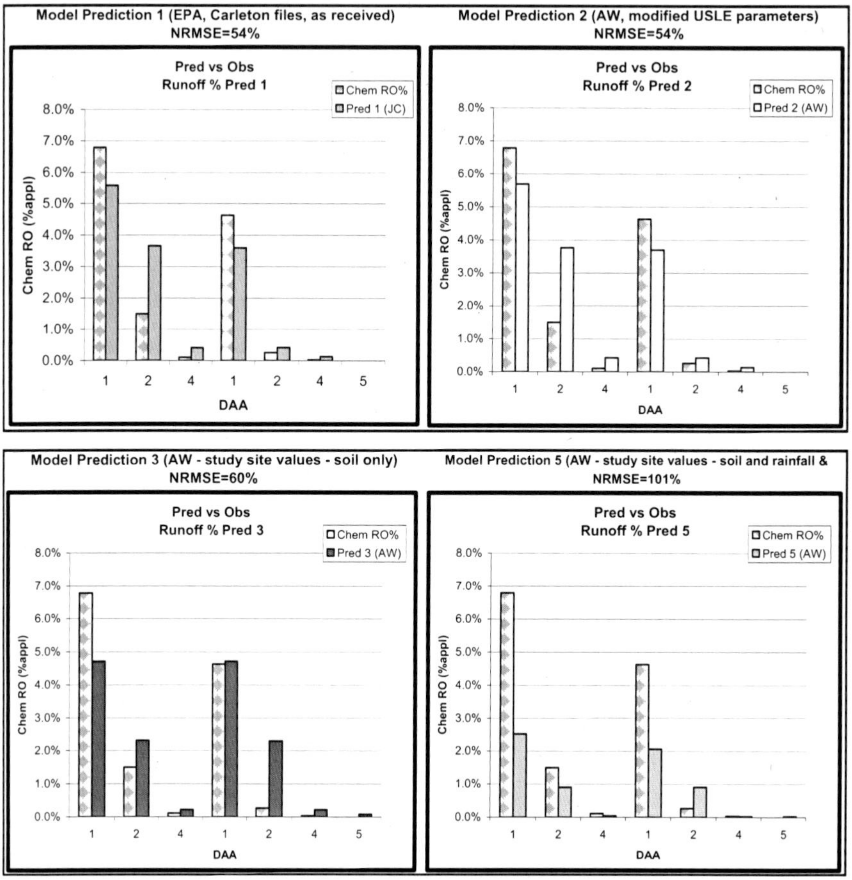

Figure 8. 2, 4-D – Evaluation of USEPA Turf Scenario model predictions of transport in runoff.

CHEM	CHLORPYRIFOS		Jun-94			Jul-94			
P/O	Source/Run	Note	1daa	2daa	4daa	1daa	2daa	4daa	5daa
Observed	Field Notes	Study File	0.05%	0.09%	0.04%	0.04%	0.04%	0.01%	0.03%
Predicted	CHPcec01	JC file	1.9%	1.5%	0.3%	2.1%	1.7%	0.4%	0.3%
Predicted	CHPcec04	AW mod	0.13%	0.19%	0.05%	0.18%	0.26%	0.07%	0.10%
Predicted	CHPcec10	AW mod2	0.14%	0.17%	0.04%	0.21%	0.24%	0.06%	0.10%
Predicted	CHPcec14	AW3 CN86	0.09%	0.11%	0.02%	0.11%	0.16%	0.02%	0.06%

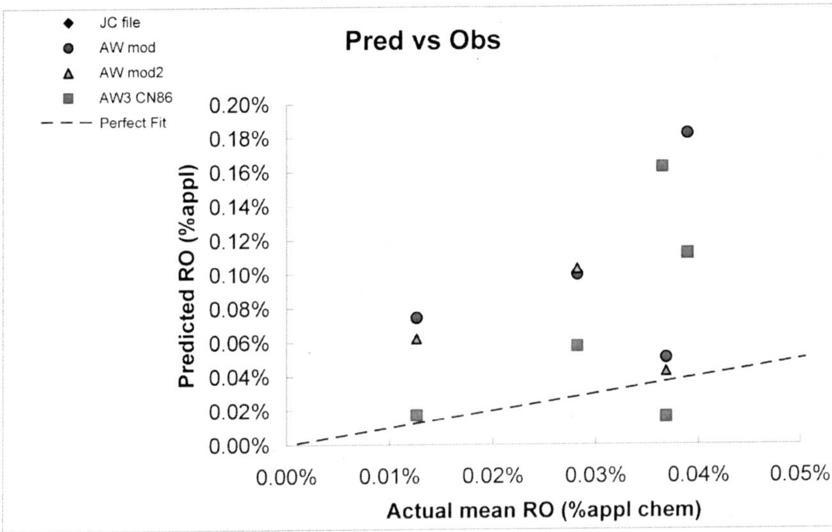

MODEL FIT		NRMSE
CHPcec01	JC file	3256 %
CHPcec04	AW mod	283 %
CHPcec10	AW mod2	281 %
CHPcec14	AW3 CN86	142 %

Observed Uses plots 6,8 not 12 as 12 has strange low runoff numbers
JC file as received
AW mod changes to USLE only
AW mod2 soil values reflect study file
AW3 CN86 soil, met (irri+rain) from study file, CN=86 to fit hydrology

Figure 9. Chlorpyrifos – Model-predicted versus observed runoff from UGA small-plot turf study.

CHLORPYRIFOS			1994	Jun-94			Jul-94			
P/O	Source/Run		Note	6/14/94	6/15/94	6/17/94	7/18/94	7/19/94	7/21/94	7/22/94
			DAA	1	2	4	1	2	4	5
Observed	Front Page	Study File	Chem RO%	0.05%	0.09%	0.04%	0.04%	0.04%	0.01%	0.03%
Predicted	CHPcec01	JC file	Pred 1 (JC)	1.93%	1.54%	0.31%	2.09%	1.68%	0.35%	0.35%
Predicted	CHPcec04	AW mod	Pred 2 (AW)	0.13%	0.19%	0.05%	0.18%	0.26%	0.07%	0.10%
Predicted	CHPcec10	AW mod2	Pred 3 (AW)	0.14%	0.17%	0.04%	0.21%	0.24%	0.06%	0.10%
Predicted	CHPcec14	CN86	Pred 5 (AW)	0.09%	0.11%	0.02%	0.05%	0.16%	0.02%	0.06%

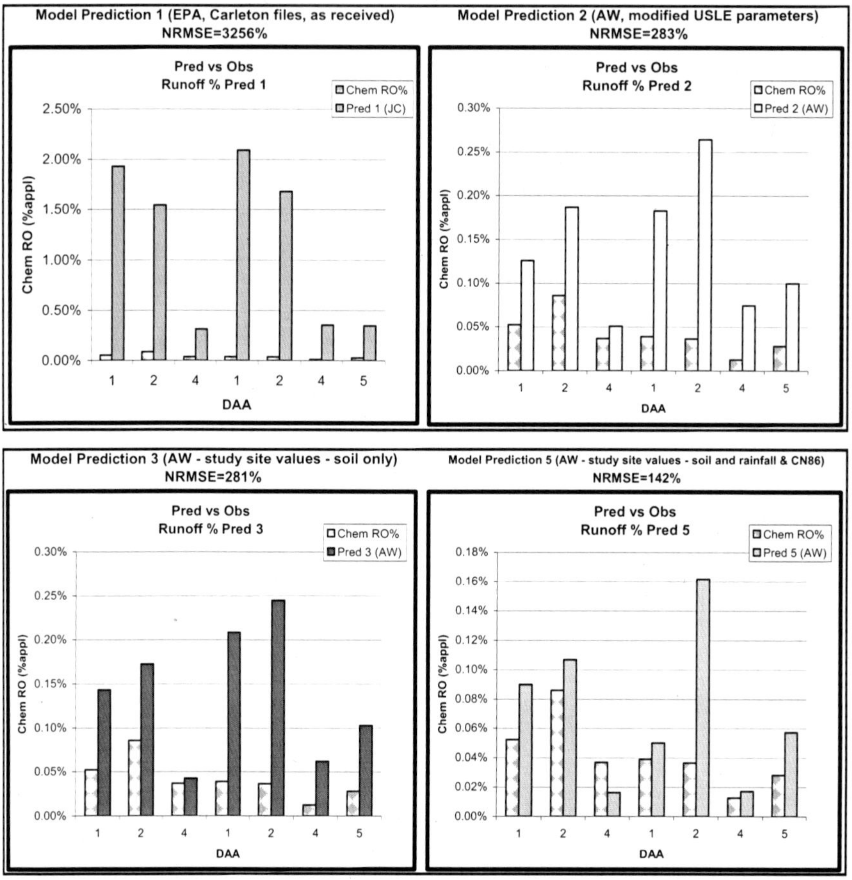

Figure 10. Chlorpyrifos – Evaluation of USEPA Turf Scenario using model predictions of transport in runoff.

CHEM	DITHIOPYR		Aug-93					Oct-93				
P/O	Source/R	Note	1daa	2daa	4daa	8daa	11daa	1daa	2daa	4daa	8daa	11daa
Observed	Hong Smith	Tbl 10,11	0.68%	0.65%	0.14%	0.05%		0.42%	0.41%	0.16%	0.09%	0.46%
Predicted	DICcec01	JC file	1.97%	1.71%	0.39%	0.21%	0.56%	0.32%	0.76%	0.24%	0.19%	0.54%
Predicted	DICcec04	AW mod	0.48%	0.70%	0.21%	0.14%	0.45%	0.33%	0.82%	0.26%	0.20%	0.58%
Predicted	DICcec10	AW mod2	0.56%	0.65%	0.18%	0.14%	0.42%	0.38%	0.74%	0.22%	0.17%	0.52%
Predicted	DICcec14	AW3 CN86	0.30%	0.39%	0.06%	0.04%	0.22%	0.17%	0.44%	0.08%	0.05%	0.28%

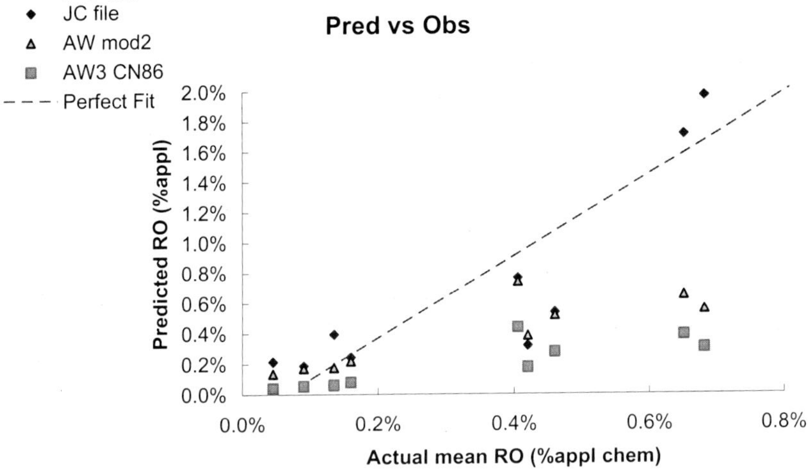

MODEL FIT		NRMSE
DICcec01	JC file	171 %
DICcec04	AW mod	65 %
DICcec10	AW mod2	54 %
DICcec14	AW3 CN86	57 %

JC file as received
AW mod changes to USLE only
AW mod2 soil values reflect study file
AW3 CN86 soil, CN=86

OBSERVED DATA (Hong and Smith, tables 10,11)

	Aug			Oct		
DAT	Granule	EC	Mean	Granule	EC	Mean
1	0.48%	0.88%	0.68%	0.20%	0.64%	0.42%
2	0.49%	0.81%	0.65%	0.30%	0.51%	0.41%
4	0.09%	0.18%	0.14%	0.10%	0.22%	0.16%
8	0.02%	0.07%	0.05%	0.03%	0.15%	0.09%
11	nr	nr	-	0.08%	0.46%	0.46%

nr, not reported

Figure 11. Dithiopyr – Model-predicted versus observed runoff from UGA small-plot turf study.

DITHIOPYR			1993	Aug-93					Oct-93				
P/O	Source/Run		Note	8/23/93	8/24/93	8/26/93	8/30/93	9/2/93	10/24/93	10/25/93	10/27/93	10/31/93	11/3/93
			DAA	1	2	4	8	11	1	2	4	8	11
Observed	Hong Smith	Tbl 10,11	Chem RO%	0.68%	0.65%	0.14%	0.05%	nr	0.42%	0.41%	0.16%	0.09%	0.46%
Predicted	DICcec01	JC file	Pred 1 (JC)	1.97%	1.71%	0.39%	0.21%	0.56%	0.32%	0.76%	0.24%	0.19%	0.54%
Predicted	DICcec04	AW mod	Pred 2 (AW)	0.48%	0.70%	0.21%	0.14%	0.45%	0.33%	0.82%	0.26%	0.20%	0.58%
Predicted	DICcec10	AW mod2	Pred 3 (AW)	0.56%	0.65%	0.18%	0.14%	0.42%	0.38%	0.74%	0.22%	0.17%	0.52%
Predicted	DICcec14	CN86	Pred 5 (AW)	0.30%	0.39%	0.06%	0.04%	0.22%	0.17%	0.44%	0.08%	0.05%	0.28%

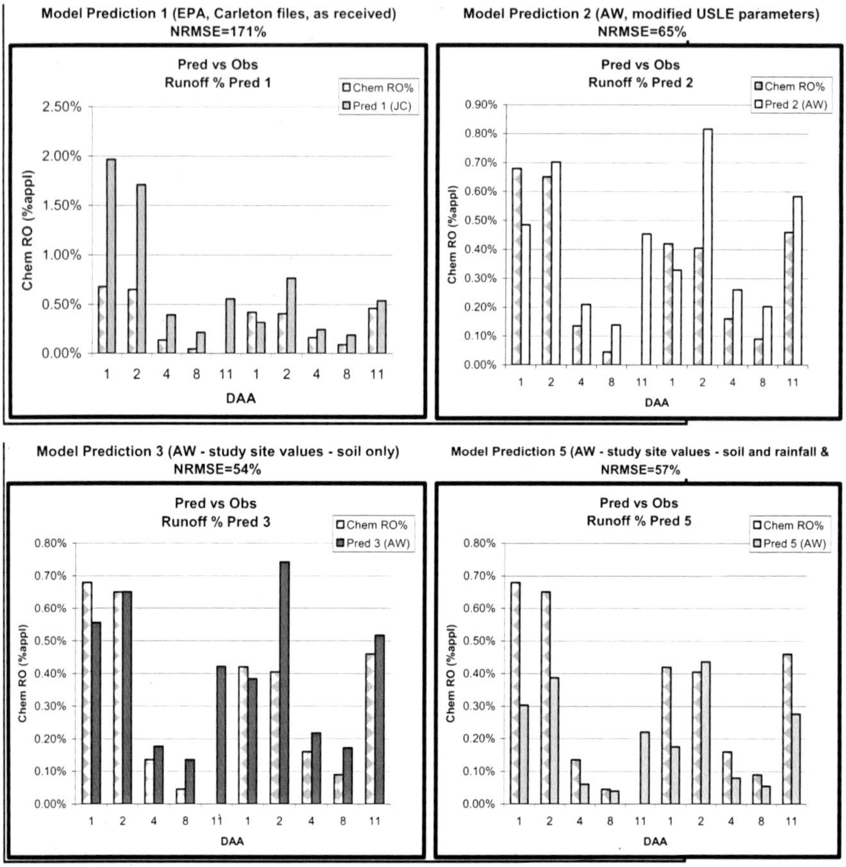

Figure 12. Dithiopyr – Evaluation of USEPA Turf Scenario using model predictions of transport in runoff.

In the first iteration of examination and adjustment of the model parameters, the authors focused on soil erosion and plot area. These parameters were adjusted to better reflect the expected Universal Soil Loss Equation (USLE) parameter estimates for sodded or grassed areas, and the particular soil in the UGA study plots. Crop cover was modeled at 100% (full grass cover is expected to occur quickly after establishment in well managed turf), the USLE K-factor (organic matter content) was adjusted from 0.32 to 0.15, the USLE LS-factor (slope length) was adjusted from 0.79 to 0.27 to better reflect a 25' long 5% slope, per PRZM manual recommendations; and finally, the USLE C factor (cover management factor) was changed from 0.13 to 0.004 (highly productive meadow, as a surrogate for well managed turf).

This iteration produced little change to the USEPA work group predictions for transport of mecoprop, dicamba and 2 ,4-D, and this result was consistent with expectations for agrochemicals that would primarily be transported dissolved in runoff. The changes in USLE factors improved the results for chlorpyrifos and dithiopyr somewhat, but their transport was still significantly over-predicted. The reduction of the over-prediction from twenty times to a maximum of five times above measured was attributed to the much more reasonable soil erosion values, which were reduced from gram per liter values to milligram per liter values in the runoff stream.

Iteration 2 concentrated on matching the site specific soil measurements as reported by Ma et al (*9*). Values for Soil Horizon 1 (the thatch layer) and Soil Horizon 2 (the first true soil layer) were adjusted. PRZM uses a tipping bucket hydrology model for directing water flow down the soil profile, and correct selection of the field capacity and wilting point moisture contents is necessary to accurately model soil moisture content and wetting/drying cycles.

The results of this iteration showed good agreement between measured and predicted values for mecoprop, dicamba and 2, 4-D, if a little under-predicted in the case of mecoprop. Much better agreement between measured and predicted values for the more strongly bound chemicals was achieved, with most runoff events for dithiopyr and chlorpyrifos showing agreement within a factor of two, albeit these discrepancies were still over-predictions.

The change of soil type and properties should have resulted in a change in antecedent soil moisture conditions. An examination of predicted runoff confirmed that runoff was, in this iteration, over-predicted. This suggested a re-examination of the selected curve number, which was especially justified in that plot-specific and event runoff volumes were available.

Iteration 3 re-adjusted the curve number sequentially, with the best fit (normalized root mean square error of 30%) obtained with a curve number of 86. A lower curve number, although enabling better predictions of water runoff, would also change chemical transport predictions. Examination showed that losses of mecoprop, dicamba and 2, 4-D were now under-predicted; predictions for dithiopyr and chlorpyrifos were much improved, however, especially for the latter.

Conclusions

Our review of the USEPA Turf Scenario, using the data collected from the University of Georgia turf plots, showed that plot to plot variability was important in interpreting the results, and modeling the studies. This was considered in our re-examination, and is deserving of further investigation. Runoff events generated at different times, even using similar equipment and with the best intention of repeatability, will differ because of the inherent variabilities of soil type, plot size, grass cover, thatch development, and especially in soil moisture content and time to runoff. Consideration of sediment erosion, and the potential for eroded soil to transport pesticides, is important, and the inclusion of moderately to strongly adsorbed compounds in the original UGA studies showed this very clearly. The UGA research also demonstrated that well-managed turf has a low potential for soil erosion, and that adsorbed chemicals are not transported from turf in significant amounts. The approach adopted by the USEPA turf work group to model turf with an increased organic matter content in a modified surface layer seems appropriate. It is suggested that the degree to which the soil surface layer's organic matter value should be modified may vary from turf to turf, depending on the quality of the turf management and the degree of thatch build up.

Acknowledgements

The authors wish to thank Professor Emeritus Albert Smith and colleagues at the University of Georgia for conducting such a comprehensive study, for including such a wide range of interesting chemicals, and for allowing us access to the study files. The authors also thank the United States Golf Association for proactively enabling the UGA research, and for allowing us access to the previously unpublished summaries of these studies. The authors additionally wish to thank the USEPA turf work group for the provision of the original PRZM input files, and BASF, Bayer, Chemtura, Dow, DuPont, Syngenta, and Valent for their support of this work.

References

1. Carleton, J., Lin, J. and Corbin, M – The EFED turf modeling workgroup. *Development of a Modeling Approach to Estimate Runoff of Pesticide Residues from Managed Turf Grass.* 2002. Informal paper presented to USEPA Exposure Modeling Work Group (EMWG).
2. Carsel, R.F., C.N. Smith, L.A. Mulkey, J.D. Dean, and P. Jowise. 1984. User's *Manual for the Pesticide Root Zone Model (PRZM): Release 1.* U.S. Environmental Protection Agency, Athens, GA. 10-3 EPA-600/3-84-109.
3. Smith, A.E. and Bridges, D.C. Movement of Certain Herbicides Following Application to Simulated Golf Course Greens and Fairways. *Crop Science.* **1996**, 26(6):1439-1445.
4. Hong, S. and Smith, A.E. Potential Movement of Dithiopyr Following Application to Golf Courses. *Journal of Environmental Quality.* **1997**, 26:379-386.
5. Mullins, J.A., R.F. Carsel, J.E. Scarbrough, A.M. Ivery. *PRZM-2, A Model for Predicting Pesticide Fate in the Crop Root and Unsaturated Soil Zones: Users Manual for Release 2.0.* 1993, U.S. Environmental Protection Agency, Environmental Research Laboratory, Athens, GA.
6. Carsel R.F., Imhoff J.C., Hummel P.R., Cheplick J.M., and Donigian A.S., Jr. *PRZM-3, A Model for Predicting Pesticide and Nitrogen Fate in the Crop Root and Unsaturated Soil Zones: Users Manual for Release 3.12.* Center for Exposure Assessment Modeling (CEAM) National Exposure

Research Laboratory - Ecosystems Research Division Office of Research and Development (ORD) U.S. Environmental Protection Agency (U.S. EPA).

7. Hurto, K.A., Turgeon, A.J. and Spooner, L.A. Physical Characteristics of Thatch as a Turfgrass Growing Medium. *Agronomy Journal*. **1980**, 72:165-167.
8. Armbrust, K. and H.B. Peeler. Effects of Formulation on the Run-off of Imidacloprid from Turf. *Pesticide Management Science*. **1999**, 58:7, 702-706.
9. Ma, Q., Smith, A.E., Hook, J.E., Bridges, D.C. Surface Transport of 2, 4-D from Small Turf Plots: Observations Compared with GLEAMS and PRZM-2 Model Simulations. *Pesticide Science*. **1999**, 55:4, 423-433.

Chapter 11

Modeling 2,4-D Transport in Turfgrass, Thatch and Soil

S. Raturi, R.L. Hill* and M.J. Carroll

Department of Natural ResourceSciences and Landscape Architecture University of Maryland, College Park, MD 20742

The transport of 2,4-D [(*2,4-dichlorophenoxy) acetic acid*] was measured for replicated soil columns containing a surface layer of turfgrass thatch and for soil columns devoid of thatch. Following the application of bromide to determine transport parameters, 2,4-D was surface-applied to undisturbed columns under steady state unsaturated conditions. Linear equilibrium (LEM), two-site non-equilibrium (2SNE) and one-site kinetic non-equilibrium (1SNE) models were curve-fitted to experimentally determined breakthrough curves. Modeling of bromide transport did not present strong evidence of significant two domain flow. All models provided reasonable estimates of 2,4-D transport, with slightly improved fits from the 2SNE model when the retardation factor was a fitting parameter. When retardation factors based on laboratory-measured adsorption coefficients were used, significantly improved fits from the 2SNE model were obtained in comparison to the LEM and 1SNE models, suggesting the occurrence of both instantaneous and kinetically driven adsorption. Parameter estimations of 2,4-D retardation factors based solely on curve fitting techniques may result in inappropriate model selection, although excellent curve fit solutions during model calibration may have been previously obtained.

Introduction

Evaluation of the potential for pesticides to contaminate groundwater requires an understanding of the transport mechanisms that occur in the field to accurately represent these mechanisms in simulation models. The presence of thatch complicates prediction of pesticide transport in turf systems, since surface-applied pesticides must pass through an organically rich thatch layer prior to entering the soil. The linear equilibrium (LEM), the two-site kinetic non-equilibrium (2SNE), and the one-site kinetic non-equilibrium (1SNE) models are process based forms of the convective dispersive equation (CDE) used to describe pesticide transport within soils. Alternate forms of this equation may be used to describe single or two domain physical flow phenomena or to describe the contribution of instantaneous or kinetically driven adsorption during the transport process.

Thatch has a pore space arrangement similar to that of a course sand and a chemical composition resembling a young organic soil (*1*). The high organic matter content of thatch allows this medium to readily sorb non-polar compounds (*2, 3, 4*). The rapid drainage properties of this medium result in short solution residence times, which can minimize the sorption of ionic compounds to thatch (*5, 6*). This suggests that solute transport models which use non-equilibrium or two-site sorption may be better able to predict pesticide transport in thatched turf than an LEM.

Convective dispersive equation based models use a retardation factor (R) to account for pesticide sorption. Several approaches have been used to obtain R for model simulations (*7, 8, 9*); most, however, involve some form of curve fitting to the data being modeled. Obtaining R in such a manner invariably improves model performance, but leaves open to interpretation the true nature of pesticide transport sorption dynamics. Retardation factors based on independently measured sorption coefficients using modified batch flow techniques should be more appropriate than using retardation factors derived as simple optimized fitting parameters, since there is a physical basis for the retardation factor which is based on the experimental conditions.

The objectives of this study were (1) to evaluate the effects of thatch on 2,4-D transport through undisturbed soil columns, (2) to compare the use of LEM, 2SNE and 1SNE models to predict 2,4-D transport through the soil columns containing a surface layer of thatch and columns devoid of thatch and (3) to evaluate the effectiveness of retardation factors based on laboratory measured adsorption coefficients and model-fitted retardation factors to simulate 2,4-D transport.

Theoretical Background

The one dimensional convective dispersive equation (CDE) for steady-state transport of a solute through homogeneous soil is (*10*):

$$R*\delta C/\delta t = D*(\delta^2 C/\delta x^2) - v* (\delta C/\delta x)$$

Where C is solution phase solute concentration (μg cm^{-3}), t is time (h), D is the hydrodynamic dispersion coefficient (cm^2 h^{-1}), R is the retardation factor (dimensionless), x is distance from solute origin (cm) and v is the average pore water velocity (cm h^{-1}). The R term reduces to one for non-reactive solutes and is greater than 1 when solute retention occurs. The retardation factor is defined as (*11*):

$$R = 1+[\rho K_f\,(1/n)C^{\,(1/n-1)}\,/\theta$$

where ρ is the soil bulk density (g cm^{-3}), θ is the volumetric water content (cm^3 cm^{-3}) and K_f and 1/n are Freundlich empirical distribution coefficient constants that characterize sorption.

The simplest approach is to assume that all pesticide sorption sites are identical, and that equilibrium occurs instantaneously between the pesticide in the bulk soil solution and the pesticide adsorbed. This mathematical approach is called linear equilibrium sorption. Where bimodal porosity leads to two-region flow, or situations where the sorption process is controlled by two-site kinetic non-equilibrium sorption processes, non-equilibrium models may more accurately describe the transport of pesticides through soil. Chemical non-equilibrium models consider adsorption on some of the sorption sites to be instantaneous, while sorption on the remaining sites is governed by first order kinetics (*12*). The two-site chemical non-equilibrium model (2SNE) conceptually divides the porous medium into two sorption sites: type-1 sites assume equilibrium sorption and type-2 sites assume sorption processes as a first-order kinetic reaction (*13*). In contrast, physical non-equilibrium is often modeled by using a two-region dual porosity type formulation. The two-region transport model assumes the liquid phase can be partitioned into mobile (flowing) and immobile (stagnant) regions. Solute exchange between the two liquid regions is modeled as a first-order process. The concepts are different for both chemical and physical non-equilibrium CDE, however, they can be put into the same dimensionless form (2SNE) for conditions of linear adsorption and steady-state water flow (*14*):

$$\beta * R * \delta C_1/\delta t = 1/P * (\delta^2 C_1/\delta x_2) - \delta C_1/\delta x - \omega(C_1 - C_2) - \mu_1 C_1 + \gamma_1(x)$$

$$(1-\beta) * R * \delta C_2/\delta t = \omega(C_1 - C_2) - \mu_2 C_2 + \gamma_2(x)$$

Where the subscripts 1 and 2 refer to equilibrium and non-equilibrium sites, respectively, β is a partitioning coefficient, ω is a dimensionless mass transfer coefficient, P is the Peclet number and μ (h^{-1}) and γ (ug h^{-1}) define first-order decay and zero-order production terms, respectively, each represented in component contributions of both the liquid and solid phases.

Customarily, β and ω are obtained by fitting solute BTC's to the non-equilibrium model using a non-linear least squares minimization technique (*15*). The values of β and ω obtained from the BTC's of non-interacting solutes can be used to evaluate the potential contributions from two-region flow. In the absence of two-region flow, β and ω may be used to evaluate the contributions from two-site kinetic non-equilibrium sorption (*16*). For interacting solutes, β represents the fraction of instantaneous solute retardation in the two-site non-equilibrium model, and ω the ratio of hydrodynamic residence time to characteristic time for sorption. They are equivalent to $\beta = (\theta + f\rho K)/(\theta + \rho K)$ and $\omega = k_2(1-\beta)RL/v$, where f is the fraction of equilibrium-type sorption sites, K is K_f when 1/n is unity, L is the length of transport (cm) and k_2 is the desorption rate constant (h^{-1}).

The one-site non-equilibrium model is a special case of the two-site non-equilibrium model. A one-site model assumes that sorption of the pesticide is kinetically driven (type-2 sites), thus the fraction of type-1 sites (f) is reduced to zero (*17*). The dimensionless coefficients used for the one-site model are the same as those used for the two-site model except that β is defined as $\beta = 1/R$, and ω as $\omega = k_2(R-1)L/v$ (17).

Materials and Methods

Sample Collection

Soil and turfgrass thatch were collected from two sites at the University of Maryland Turfgrass Research and Education Facility in Silver Spring, Maryland. One site was a three and half year old stand of Southshore creeping bentgrass (*Agrostis stolonifera*), and the other a six year old stand of Meyer Zoysiagrass (*Zoysia japonica* **Steud**.). Visual inspection of the bentgrass site revealed the presence of a finely granulated 1.5 to 2.0 cm thick thatch layer. The zoysiagrass site contained a 3.0 to 3.5 cm thick thatch layer that consisted primarily of non-decomposed and partially decomposed rhizomes, stolons and tillers. The soil at the zoysiagrass site was classified as a Sassafras loamy sand (fine loamy, mixed, mesic, Typic Hapludult; 81.2% sand, 10.2% silt, and 8.7% clay) whereas the soil at the bentgrass site was classified as a Sassafras sandy loam (fine sandy, mixed, mesic, Typic Hapludult; 71.2% sand, 15.8% silt, and 12.8% clay). The saturated soil hydraulic conductivity was 24.4 cm h^{-1} at the zoysiagrass site and 18.2 cm h^{-1} at the bentgrass site.

The thatch and soil used to determine 2,4-D sorption isotherms were collected by removing the thatch using a sod cutter. Prior to using the sod cutter, all verdure was removed by scalping the turf with a walk-behind greens mower. The intact rolls of the turfgrass thatch were shredded using a modified wood chipper and the shredded field moist material passed through a 4 mm screen. The soil directly beneath the thatch (2 cm depth) was collected using a shovel, and the field moist soil passed through a 4 mm screen. The soil columns used in the leaching study were extracted from the surface of each site using a specially designed drop hammer-sleeve assembly. Four soil and four soil plus thatch columns, 12 cm length by 10 cm diameter, were collected from each site. The columns containing soil only were obtained after using a shovel to remove all above ground thatch and foliage. The columns were brought to the laboratory and saturated from the bottom immediately after collection.

Sorption Isotherms

A modified batch/flow technique was used to measure pesticide sorption (*5*). The technique involved the use of a mechanical vacuum extractor. This device controls the rate at which a solution moves through a column of thatch or soil. The columns were created by packing known amounts of media into syringe tube barrels after placing a sheet of glass fiber filter paper (Fisher Scientific, Pittsburgh, PA, Cat. No: 09-804-70C) into the bottom of each barrel. Since the sample was not shaken during the procedure, little disruption of the medium aggregates and organic matter occurred. Moreover, the flowing conditions used in this modified batch/flow technique better represent the physiochemical interactions that occur in the field.

A combination of technical grade 2,4-D and ring-labeled ^{14}C 2,4-D were used to determine the sorption of 2,4-D to thatch and soil. Sorption isotherms were determined by leaching 30 mL of 1, 10, 30, or 100 mg 2,4-D L^{-1} through samples of thatch and soil for 24 hours. All four solutions contained 2.31×10^5 Bq L^{-1}, of ^{14}C 2,4-D. The radioactivity of 1 mL of leachate plus 5 mL of scintillation cocktail was determined by liquid scintillation counting (LSC). Sorption of 2,4-D to any material other than thatch or soil was accounted for by including syringe tube blanks. The blanks were identical to the syringe tubes containing thatch or soil, except they contained no thatch or soil. Sorption of 2,4-D at 24 h was fitted to the linear form of the Freundlich equation. Regression analyses were used to calculate the capacity (K_f) and intensity (1/n) of sorption to each medium. Student's t-tests were used to test for homogeneity of slopes and to compare equation intercepts.

Leaching Experiment

The bottom end of the columns obtained from the field were placed into separate funnels fitted with a rubber o-ring and a 12-μm pore diameter saturated, porous, stainless steel plate. The columns were made vacuum tight to each funnel using adhesive acrylic caulking and the funnel inserted into one port of a multi-port vacuum chamber. A null balance vacuum regulator was used to maintain a constant pressure of -10 kPa within each vacuum chamber.

A 0.001 M $CaCl_2$ solution was continuously applied to each column using a specially designed drop emitter that uniformly distributed the solution to the surface of each column (modified design of *18*). Leachate was collected in sterile plastic cups beneath the funnel of each column within the vacuum chamber. Once steady-state flow (0.85 cm h^{-1}) was achieved in all columns, 10 mL of 300 mg bromide L^{-1} (KBr salt) was surface-applied uniformly to each column. Leachate was then collected every 30 minutes for the next 12 hours. The concentration of Br^- in the leachate was determined using standard ion chromatography techniques outlined in the Dionex Users Guide.

After the initial leaching period, 10 mL of 88 mg 2,4-D L^{-1} was uniformly surface-applied to each column. The 2,4-D solution contained 2.31×10^5 Bq L^{-1} of ring-labeled ^{14}C 2,4-D. The addition to each column was equivalent to a field rate of 1.12 kg 2,4-D ha^{-1}. After adding 2,4-D, the leaching solution inputs and vacuum applied to the base of each column were discontinued for 24 hours to permit sorption of 2,4-D to the thatch and soil. During this time, all columns were covered with plastic wrap to prevent volatile losses of 2,4-D. After the 24 h adsorption period, the plastic wrap was removed, the emitters placed back atop each column, and the vacuum engaged. Leachate was then collected every half an hour for the next 18 h with 2,4-D in the leachate being determined by LSC as previously described. To verify that the radioactivity measured by LSC was ^{14}C 2,4-D and not one of its primary metabolites, every 4 h during the leaching event 1mL subsamples of leachate were collected from a single column for each of the four column treatments. A 25 cm x 4.6 mm ID 5μm Supelcosil LC-18 (Supelco p/n 5-8298, Bellefonte, PA) column installed into an Hewlett Packard (HP) model 1050 liquid chromatographic system equipped with a quartenary pumping module, automatic liquid sampler and HP model 79853A variable wavelength UV detector was used to determine the concentration of 2,4-D in these samples. Analytical standards of 2,4-D were analyzed concurrently with the leachate samples to confirm the accuracy of the high performance liquid chromatography (HPLC) analysis. The limit of quantification was 0.02 μg 2,4-D mL^{-1}.

After collecting the last leachate sample, the columns were removed from the vacuum chamber and sectioned into halves. In the columns containing thatch, the thickness of the thatch layer was measured before separating the thatch and soil. One half of each core section was used to determine the water content in the section. The other half of the section was immediately frozen for later determination of the amount of 2,4-D present in the section. At a later date, the frozen section was thawed and shaken for 2 hours in a 50:50 water and methanol solution. The resulting slurry was then subjected to vacuum filtration and the filtrate analyzed for ^{14}C. The amount of ^{14}C remaining in the sample was determined through combustion using a biological material oxidizer with the amount of ^{14}C evolved measured by LSC.

Estimating Transport Parameters From Breakthrough Curves (BTC's)

Convective transport parameters were estimated by a least squares minimization procedure (CXTFIT, *19*) using the bromide breakthrough data. All CXTFIT calculations were performed under flux type boundary conditions. Actual mean pore water velocities were used and the retardation factor, R, was assumed to be equal to 1. One and two domain flow forms of the convective dispersive equation were curve-fit to the bromide leachate data. Values of the dispersion coefficient were used in subsequent 2,4-D simulations.

The LEM model was fitted to the 2,4-D transport data using CXTFIT. The 2SNE and 1SNE models were fitted using CXT4 (*13*). All models used calculated mean pore water velocities and the bromide-fitted dispersion coefficients. Degradation coefficients (μ) for the LEM model were estimated by applying an exponential decay function to mass balance quantities. The dimensional degradation coefficients ($\Psi = \mu L/v$; *20*) in the 2SNE and 1SNE models were calculated from the appropriate μ values. The retardation factors were calculated based on the column measured values of θ, ρ, maximum pesticide breakthrough concentrations and 2,4-D adsorption coefficients (K_f and

1/n). Pesticide retardation factors for individual columns were calculated using thatch and soil K_fs in a volume-averaged approach where the relative length of the thatch and soil layers were used as weighing factors in calculating a mean retardation factor for each column. Pulse is the duration of solute addition and was a fitting parameter during all model simulations. The dimensionless partitioning coefficient (β), and the dimensionless rate coefficient (ω) which specify the degree of either chemical or physical non-equilibrium were fitted for the 2SNE model. For the 1SNE model, the value of β was calculated as $\beta = 1/R$. The value of ω and pulse were fitted in the 1SNE model.

Simulations were repeated for all columns a second time, with retardation factors being fitted so comparisons could be made of model fits using measured and fitted retardation factors. The value of R determined from the two-site model was assumed to be the same for the 1SNE model. The value of β was then calculated as $\beta = 1/R$. The value of ω and pulse were fitted in the 1SNE model simulations. Simulations were repeated for all columns a third and fourth time, and the previously stated methods were followed in each case except the degradation term was assumed to be zero.

Results and Discussion

Freundlich sorption parameters for 2,4-D in thatch and soil are presented in Table I. The sorption of 2,4-D to thatch was greater than to soil. The 2,4-D sorption capacity of the two turfgrass species were similar, and are represented by a single set of Freundlich parameters.

Table I. Freundlich Sorption Parameters for 2,4-D in Bentgrass and Zoysiagrass Thatch and the Soil Residing Below Each Thatch Layer

Media	*log K_f*	*K_f*[a]	*1/n*	*r^2*
Thatch[b]	0.50 (± 0.02)[c]	3.14x[d]	0.86 (± 0.01)x	0.99
Bentgrass Site Soil	-0.15 (± 0.04)	0.71y	0.86 (± 0.03)x	0.98
Zoysiagrass Site Soil	-0.46 (± 0.06)	0.35z	0.83 (± 0.04)x	0.96

[a] $K_f = mg^{(1-1/n)} L^{1/n} kg^{-1}$

[b] There was no difference in the 2,4-D sorption capacity of the two turfgrass species thatch thus a single Freundlich isotherm was used to describe the sorption of 2,4-D to turfgrass species thatch.

[c] Values in the parenthesis indicate standard errors of estimates.

[d] Values followed by the same letter in a column are not significantly different ($P \leq 0.05$).

Leaching

The physical properties and transport conditions for the soil and thatch plus soil columns are summarized in Table II. The soil water contents were slightly greater for the columns containing a surface layer of thatch compared to the columns devoid of thatch. The presence of thatch also decreased the mean bulk density of the thatch plus soil columns compared to the columns devoid of thatch.

The amount of 2,4-D determined by LSC analysis was plotted against the amount determined by HPLC analysis for the columns where analysis by both analytical techniques took place. There was good agreement between the two techniques (r^2 >0.95) indicating that the LSC data represented the presence of ^{14}C 2,4-D and not one or more of 2,4-D's metabolites. Total mass LSC recoveries of 2,4-D within the columns were 81.18 (±2.35)% for the zoysiagrass thatch+soil, 72.26 (±3.02)% for the zoysiagrass soil, 84.91 (±8.17)% for the bentgrass thatch + soil, and 90.86 (±7.03)% for the bentgrass soil. The bentgrass columns devoid of thatch had the greatest 2,4-D leaching losses (43.11±1.10%). Conversely, columns having a surface layer of bentgrass thatch had the lowest 2,4-D leaching losses for the four column types examined (17.45 ± 1.83%). Columns having a 3.5 year old, 1.7 cm surface layer of bentgrass thatch were more effective (P=0.0078) in reducing 2,4-D leaching than columns having a 6 year old, 3.2-cm surface layer of zoysiagrass thatch (29.03 ± 3.01%). There was no difference (P=0.162) in the amount of 2,4-D leached from the zoysiagrass columns devoid of thatch (34.35 ± 2.04%) than from the zoysiagrass columns containing thatch. The coarser nature of the zoysiagrass thatch may have limited the amount of 2,4-D that was initially intercepted, compared to the more tightly intertwined bentgrass thatch. The leaching results demonstrate that the turfgrass species from which a mature thatch layer originates can have a greater influence on 2,4-D attenuation than does the age or thickness of the thatch layer.

Model Evaluation

Both one and two domain flow models did well in estimating bromide transport with reasonable estimations of transport parameters obtained for all columns. Peak bromide concentrations occurred prior to the leaching of one pore volume in some columns which might suggest two domain flow, but both models performed well and gave close fits of the peak concentrations with r^2 values of 0.98 to 0.99 obtained for most columns. Since both the one and two domain

Table II. Mean Physical Properties and Experimental Parameters for Columns Containing a Surface Layer of Thatch and Columns Devoid of Thatch Used in the 2,4-D Leaching Study

Column ID	*Mean Pore Water Velocity* $cm\ h^{-1}$	*Darcy Flux* $cm\ h^{-1}$	*Soil Water Content* $cm^3 cm^{-3}$	*Bulk Density* $g\ cm^{-3}$
Zoysiagrass Thatch+Soil	2.41	0.86	0.34	1.30
Zoysiagrass Site Soil	2.60	0.83	0.32	1.66
Bentgrass Thatch+Soil	2.77	0.87	0.31	1.24
Bentgrass Site Soil	3.43	0.88	0.25	1.54

flow models resulted in good fits to the measured bromide leaching, there is not strong evidence that significant amounts of two domain flow was occurring.

Mass balance derived 2,4-D degradation coefficients (μ) values for columns containing the zoysiagrass thatch and bentgrass thatch were 0.014 and 0.012 h^{-1}, respectively. Mass balance derived 2,4-D μ values for zoysiagrass and bentgrass site soil columns were 0.034 and 0.007 h^{-1}, respectively. When μ and Ψ ($\Psi = \mu L/v$) for the appropriate models were included in the model simulations, there were no perceivable improvements in the quality of the fits compared to model simulations where a degradation term was not included. The solute leaching concentrations comparing the liquid scintillation counting and HPLC methodologies also indicated that there was minimal degradation of 2,4-D in the leachate. Thus, model simulation values without the use of degradation coefficients are presented and discussed (Tables III and IV).

If model evaluation is based on the coefficient of determination, all three models described 2,4-D transport fairly well, with slightly improved fits resulting from the 2SNE model when R was a fitting parameter. This overall quality of fits may be partially attributable to volume averaging thatch physical properties over the column length when calculating transport parameters. The Damköhler number (ω), which is a ratio of hydraulic residence time to reaction time and, as such, characterizes the degree of non-equilibrium, is often used as a criterion for linear equilibrium model validity (*21, 22*). Results of the sensitivity analysis showed that values of ω >100 were generally indicative of LEM validity and lack of significant transport non-equilibrium (*16, 23*). The fact that ω values were greater than 100 for columns containing the surface layer of thatch suggested that the LEM model may have been appropriate for describing 2,4-D transport. However, when R was based on a laboratory measured sorption coefficient, the 2SNE model gave significantly improved fits, indicating two-site non-equilibrium adsorption may have occurred. Similar results of two-site sorption non-equilibrium exhibited by 2,4-D were also reported by Khan (*24*) and Rao et al., (*25*). They attributed sorption non-equilibrium of 2,4-D to the rate-limited interaction between 2,4-D and the sorbent organic matter (*24, 26*).

The fitted retardation coefficients were 37 to 75% lower than the measured retardation factors for the LEM, and 35 to 68% lower for the 2SNE models. Model fits which were based on measured R values also gave more realistic values of β than model fits where R was a fitting parameter (R_{fit}). When R was a fitting parameter with the 2SNE model, mean values of β increased for the bentgrass thatch plus soil columns from 0.41 to 0.84 (108% increase), and increased for the zoysiagrass thatch plus soil columns from 0.42 to 0.82 (96% increase). Retardation factors based on laboratory measured sorption coefficients (R_{mes}) should be more appropriate than using retardation factors derived as a simple optimized fitting parameter, since there is a physical basis for the R value which is based on the experimental conditions. Using estimated R (R_{fit}) values in prediction models may underestimate or overestimate subsequent 2,4-D transport and may not accurately represent the physical processes occurring during 2,4-D transport, since the model is optimally fitting for the parameters. Similar conclusions were also obtained by Brusseau (*27*), where the utility of fitting a model to measured laboratory and field experiment data to model evaluation and data analysis was examined. They reported that the misuse of calibration can lead to a mistaken belief that the model accurately represents the physical system which can result in a misinterpretation of the factors controlling solute transport.

Non-equilibrium parameters (β and ω) for 2,4-D transport were optimized by fitting 2,4-D BTCs to the 2SNE model using independent estimates of R and v (Tables III and IV). Because bromide exhibited no significant two-domain flow in the columns; β and ω values obtained for the 2,4-D BTCs can be interpreted primarily as sorption related non-equilibrium parameters (*16, 19*). In this case, calculated values of f (fraction of sorbent for which sorption is instantaneous) and k2 (the desorption rate constant) using β and ω terms, respectively, may be interpreted as relating to sorption non-equilibrium (2SNE) without confounding effects from two-region or transport related non-equilibrium (*28*). Values of f were 0.19, 0.14, 0.11 and 0.30 for columns containing a surface layer of bentgrass thatch, zoysiagrass thatch, bentgrass site soil columns and zoysiagrass site soil columns, respectively. These f values indicated that a significant fraction of the sorption sites did not participate in instantaneous retardation in these columns during 2,4-D transport. Values of k2 for columns containing a surface layer of bentgrass thatch, zoysiagrass thatch,

Table III. Transport Parameters for 2,4-D Breakthrough Curves From the Linear Equilibrium (LEM), Two-Site Non-Equilibrium (2SNE) and One-Site Kinetic Non-Equilibrium (1SNE) Models for Zoysiagrass Thatch+Soil Columns (ZT) and Soil Columns Devoid of Thatch (ZS) Using Fitted (ZT_{fit}, ZS_{fit}) and Measured (ZT_{mes}, ZS_{mes}) Retardation Factors

Column ID	*Model*	v	D	R_{mes}	R_{fit}	β	ω	r^2
ZT_{mes}	LEM	2.77	3.45	4.05	--	--	--	0.15
	2SNE	2.77	3.45	4.05	--	0.419	0.218	0.93
	1SNE	2.77	3.45	4.05	--	0.247	2.62	0.31
ZT_{fit}	LEM	2.77	3.45	--	1.71	--	--	0.93
	2SNE	2.77	3.45	--	1.72	0.822	1115	0.94
	1SNE	2.77	3.45	--	1.72	0.582	385	0.94
ZS_{mes}	LEM	3.43	2.25	3.19	--	--	--	0.05
	2SNE	3.43	2.25	3.19	--	0.523	0.547	0.92
	1SNE	3.43	2.25	3.19	--	0.314	2.41	0.66
ZS_{fit}	LEM	3.43	2.25	--	1.74	--	--	0.93
	2SNE	3.43	2.25	--	1.83	0.464	7.40	0.97
	1SNE	3.43	2.25	--	1.83	0.553	5.12	0.96

bentgrass site soil columns and zoysiagrass site soil columns were 0.03, 0.02, 0.06 and 0.12, respectively, indicating that there were relatively large differences in 2,4-D desorption in the columns containing a surface layer of thatch and columns devoid of thatch.

Table IV. Transport Parameters for 2,4-D Breakthrough Curves From the Linear Equilibrium (LEM), Two-Site Non-Equilibrium (2SNE) and One-Site Kinetic Non-Equilibrium (1SNE) Models for Bentgrass Thatch+Soil Columns (BT) and Soil Columns Devoid of Thatch (BS) Using Fitted (BT_{fit}, BS_{fit}) and Measured (BT_{mes}, BS_{mes}) Retardation Factors

Column ID	*Model*	v	D	R_{mes}	R_{fit}	β	ω	r^2
BT_{mes}	LEM	2.41	6.87	4.77	--	--	--	0.09
	2SNE	2.41	6.87	4.77	--	0.405	0.316	0.78
	1SNE	2.41	6.87	4.77	--	0.212	4.62	0.27
BT_{fit}	LEM	2.41	6.87	--	1.88	--	--	0.78
	2SNE	2.41	6.87	--	1.91	0.844	1810	0.81
	1SNE	2.41	6.87	--	1.91	0.528	248	0.79
BS_{mes}	LEM	2.61	4.61	4.75	--	--	--	0.35
	2SNE	2.61	4.61	4.75	--	0.299	0.951	0.89
	1SNE	2.61	4.61	4.75	--	0.214	2.26	0.49
BS_{fit}	LEM	2.61	4.61	--	1.51	--	--	0.87
	2SNE	2.61	4.61	--	1.79	0.633	4.99	0.94
	1SNE	2.61	4.61	--	1.79	0.564	24.61	0.91

Conclusions

The presence of bentgrass thatch reduced the leaching of 2,4-D applied to turfgrass. When 2,4-D breakthrough curves were fitted to the different forms of the convective dispersive equation, all models provided reasonable estimates of 2,4-D transport when the retardation factor was a fitting parameter. When retardation factors derived from laboratory determined sorption coefficients were used, significantly improved fits from the 2SNE model were obtained in comparison to the LEM and 1SNE models, indicating the occurrence of both instantaneous and kinetically driven adsorption. Retardation factors derived from laboratory determined adsorption coefficients provided a more realistic estimation of processes involved in 2,4-D transport. Mass balance-derived degradation coefficients did not result in improved model estimations, and had limited utility because of the minimal quantities of 2,4-D degradation observed. While there were differences in the amount of 2,4-D leached from columns containing the thatch and those devoid of thatch, the presence of thatch did not affect model performance. Column k2 values revealed that 2,4-D was more tightly sorbed to thatch than soil, however, the presence of thatch did not appear to alter the fraction of sorption sites associated with instantaneous sorption within the columns. The proportion of instantaneous sorption sites was relatively low in all columns, which may explain why non-equilibrium transport of 2,4-D was observed in all columns. More importantly, the results of this study showed that parameter estimations of 2,4-D retardation factors based solely on curve-fitting techniques may result in inappropriate model selection, even though excellent curve fit solutions during model calibration may have been previously obtained.

Acknowledgements

The authors express their appreciation to the United States Golf Association Green Section Research,The Maryland Turfgrass Council and Maryland Agricultural Experiment Station for providing financial assistance to conduct this research. Special thanks are extended to Dr. A. Herner (now retired) and Emy Pheil at the Environmental Chemistry Laboratory, USDA, Beltsville for their assistance during this research.

References

1. Hurto, K.A., Turgeon, A.J. and Spomer, L.A. *Agron. J.* **1980**, *72*, 165-167.
2. Hurto, K.A. and Turgeon, A.J. *Weed Sci.* **1979**, *27*, 141-146.
3. Dell, C.J., Throssell, C.S., Bischoff, M. and Turco, R.F. *J. Environ. Qual.* **1994**, *23*, 92-96.
4. Lickfeldt, D.W. and Branham, B.E. *J. Environ. Qual.* **1995**, 24, 980-985.
5. Raturi, S., Carroll, M.J., Hill, R.L., Pfeil, E. and Herner. A.E. *Intern. Turf. Soc. Res. J.* **1997**, *8*, 187-196.
6. Carroll, M.J, and R.L. Hill United State Golf Association Turfgrass and Environmental Research Onlin 2003, 2:(5) 1-9. (ISSN 1541-0277, http//www.usgatero.msu.edu. TGIF Record Number 85740).
7. Roberts, P.V., Goltz, M.N. and Mackay, D.M. *Water Resour. Res.* **1986**, *22*, 2047-2058.
8. Winters, S.L. and Lee, D.R. *Environ. Sci. Technol*, **1987**, 21, 1182-1186.
9. Widmer, S.K., Spalding, R.F. and Skopp, J. *J. Environ. Qual.* **1995**, *24*, 439-444.
10. Lapidus, L. and Amundşon, N.R. *J. Phys. Chem.* **1952**, *56*, 984-988.
11. Hashimoto, I.; Deshpande, K.B.; Thomas, H.C. *Ind Eng. Chem Fundamen.* **1964**, *3*, 213-218.
12. Selim, H.M., Davidson, J.M. and Mansell, R.S. **1976**,. p. 444-448. *In Proc. Summer Comput. Simul.* Conf., Wash DC. Simul Counc; La Jolla, CA.
13. van Genuchten, M. Th. and Wagenet R.J. *Soil Sci. Soc. Am. J.* **1989**, *53*, 1303-1310.

14. Nkedi-Kizza, P., Bigger, J.W., Selim, H.M., Van Genuchten, M.Th., Wierenga, P.J., Davidson, J.M. and Nielsen, D.R. *Water Resour. Res.* **1984**, *20*, 1123-1130.
15. Toride, N.; Leij, F.J.; van Genuchten, M. Th. *The CXTFIT code for estimating transport parameters from laboratory or field tracer experiments, Version 2.* Research Rpt. 137, U.S. Salinity Laboratory, USDA, ARS, Riverside, CA. 1995.
16. Gaber, H.M., Inskeep, W.P., Comfort, S.D. and Wraith, J.M. *Soil Sci. Soc. Am. J.* **1995**, *59*,160-67.
17. van Genuchten, M.Th. Res. Rep. no. 119. U.S. Salinity Lab., Riverside, CA. 1981a
18. Ogden, C.B., vanEs, H.M. and Schindelbeck. R.R. *Soil Sci. Soc. Am. J.* **1997**, *61*, 1041-1043.
19. Parker, J.C. and van Genuchten, M. Th. 1984, *Virginia Agric. Exp. Stn. Bull.* no. 84-3.
20. Gamerdinger, A.P, Lemley, A.T. and Wagenet, R.J. *J. Environ. Qual.* **1991**, *20*, 815-822.
21. Brusseau, M.L., Jessup, R.E. and Rao, P.S.C. *Water Resources Research.* **1989**, *25*, 1971-1988.
22. Rao, P.S.C., and Jessup, R.E.: **1983**, In *Chemical Mobility and Reactivity in Soil Systems*, P. 183-201. American Society of Agronomy, Madison, WI.
23. Bahr, J.M., and Rubin, J. *Water Resour. Res.* **1987**, *23*, 438-452.
24. Khan, S.U. **1973**, *Can. J. Soil Sci.* 53, 429-434.
25. Rao, P.S.C., Davidson, J.M. and Selim, H.M. *Soil Sci. Soc. Am. J.* **1979**, *43*, 22-28.
26. Brusseau, M.L., and Rao, P.S.C. *Crit. Rev. Environ. Control.* **1989**, *19*, 33-99.
27. Brusseau, M.L. *J. of Hydrol.* **1998**, *209*, 147-165.
28. Brusseau, M.L., Jessup, R.E. and Rao, P.S.C. *Environ. Sci. Technol.* **1991**, *25*,134-142.

Chapter 12

Regional Analyses of Pesticide Runoff from Turf

Douglas A. Haith[1], Matthew W. Duffany[2], Antoni Magri[3]

[1]Cornell University, Ithaca, NY 14853
[2]NY State Department of Environmental Conservation, Watertown, NY 13601-3787
[3]ESRI, Redlands, CA 92373-8100

Pesticide runoff loads from turf can vary dramatically with chemical properties and application regime, geographic location, irrigation rates and turf surface. Given the limited availability of field data, it is difficult to realistically consider the range of these variations in exposure assessments. The TurfPQ pesticide runoff model was combined with several other models and data bases to provide a general framework for efficient estimation of turf pesticide runoff loads on both a yearly and daily basis. The process was used to investigate differences in MCPP, fenarimol, iprodione and carbaryl runoff from fairways at four U.S. locations with widely differing climatic regions. Factors which accounted for the observed differences included pesticide properties and application amounts, irrigation applications and growing season runoff. The simulations indicated that runoff loads of a particular pesticide could vary by as much as an order of magnitude among the locations.

One of the significant difficulties in managing the environmental impacts of turf pesticide runoff is the immense variability in transport and fate characteristics. One pesticide may be easily washed from grass surfaces by small amounts of runoff while another resists movement, even with extreme storms. Some chemicals persist in the turf and soil for months while others are degraded within days or even hours. These variations are further compounded by differences in weather patterns between geographic locations. As a result, a program for controlling the runoff of one pesticide at one site is not likely to be adequate for another chemical and site.

A classic approach for elucidating such differences is through controlled field experiments. Given the large number of available turf pesticides and the many different weather regimes seen in an area as large as the continental U.S., this approach has limited practicality. Fortunately, many of its features can be duplicated in simulation experiments. Mathematical models are used to describe weather and runoff, and the effects of a variety of site conditions and management options can be efficiently evaluated. Nevertheless, simulation experiments of pesticide runoff are challenging. Available models often require many input parameters whose values are difficult to estimate. Information on the rates and timing of pesticide applications for a particular location may be particularly difficult to obtain.

The research described herein had two objectives. The first was to develop a general protocol for simulation studies of pesticide runoff from turf. The protocol is built around the TurfPQ pesticide runoff model *(1, 2)*, and USCLIMATE weather generator *(3)*, but the methods should be applicable to other models as well. The second objective was to demonstrate the protocol through a simulation experiment designed to study the regional differences in runoff of several pesticides applied to fairways.

Simulation Protocol

A simulation protocol consists of the design of the simulation experiment's scenario (pesticide selection, site description, length of simulation run), the specification of appropriate models, estimation of input parameters, and selection of methods for summarizing and interpreting results.

Scenario

Four pesticides were simulated: the herbicide MCPP (2-(2-Methyl-4-chlorophenoxy) propionic acid), two fungicides, fenarimol (α-(2-Chlorophenyl)-α-(4-chlorophenyl)-5-pyridinemethanol) and iprodione (3-(3,5-Dichlorophenyl)-N-(1-methylethyl)-2,4-dioxo-1-Imidazolidinecarboxamide), and the insecticide carbaryl (1-Naphthyl-N-methylcarbamate). The sites are identical, hypothetical golf fairways in Atlanta, GA; Fresno, CA; Madison, WI; and Olympia, WA. Weather characteristics for these sites are given in Table I. Temperatures and precipitation are 1971-2000 means *(4)*. Growing seasons are based on median freeze/frost dates *(5)*.

Table I. Weather Characteristics of Simulation Sites

Location	*Annual Temperature (°C)*	*Annual Precipitation (mm)*	*Growing Season*
Atlanta, GA	16	1290	Apr-Oct
Fresno, CA	17	270	Mar-Nov
Madison, WI	7	785	May-Sep
Olympia, WA	10	1285	May-Oct

It can be seen from Table I that the four sites have substantially different weather characteristics. Atlanta and Fresno both have warm climates, but Fresno is much drier and would require significant irrigation to maintain turf surfaces. Madison and Olympia are cooler and have shorter growing seasons than the other cities. Although Olympia's annual precipitation is comparable to Atlanta's, it is differently distributed. Atlanta precipitation is relatively uniformly distributed throughout the year, but Olympia has little growing season moisture. Unlike field experiments, simulations can be of any duration. It is typically as easy to make 500-year runs as 5-year ones. In general, runs should be long enough to provide reliable estimates of the phenomena of interest. In the current study, regional differences were evaluated by comparison of annual and monthly means and 1 in 10-year extreme events, and these variables could be reasonably estimated from 100 years of daily results. This does not imply that the experiments modeled 100 years of fairway operations. Rather, the 100-year run should be interpreted as producing 100 different estimates of one-year of pesticide runoff.

Simulation Models

The TurfPQ model was used in this study to simulate pesticide runoff. The model computes water and chemical mass balances on a one-day time step. Runoff volume is determined through a modified curve number equation. Pesticide in turf foliage and thatch is partitioned into adsorbed and dissolved components which are assumed to be decayed in a first order biodegradation process. In addition to decay, dissolved pesticide is removed from the system by runoff or leaching into the soil. Volatilization is neglected. In addition to daily precipitation and temperatures and pesticide application rates, the model requires four input parameters – biodegradation half-life, organic carbon partition coefficient, runoff curve number, and organic carbon content of the turf. In a validation study of 52 runoff events in four states involving 6 pesticides, TurfPQ explained 65% of the observed variation in pesticide runoff. Mean predicted pesticide runoff was 2.9% of application, compared to a mean observation of 2.1% (*1, 2*).

The USCLIMATE software package, which was used to generate daily weather data for the TurfPQ model, produces daily precipitation, minimum and maximum air temperatures and a solar radiation record for arbitrary user-specified locations in the continental U.S. Precipitation is based on a Markov chain of occurrence (wet/dry days) and a mixed exponential distribution for precipitation amount. Temperatures are described by an autocorrelation model conditioned on wet or dry days. The generated weather data are processed in several ways to produce the daily records of precipitation and temperatures required by TurfPQ. Solar radiation data are discarded and the software's March to April sequences are converted to January to December. Daily temperatures are obtained by averaging the minimum and maximum temperatures.

Input Data

Weather

Depending on the nature of the site, the weather records may be further modified to reflect the addition of irrigation. This would generally be the case for golf course turfs. In this study, irrigation was based on comparison of 3-day cumulative precipitation and potential evapotranspiration during the growing season. Whenever the 3-day precipitation is exceeded by 3-day potential evapotranspiration as computed by the Hamon equation *(6)*, irrigation is added to make up the deficit. This produces a new weather record in which precipitation entries for any day are replaced by precipitation plus irrigation.

Turf Properties

Turf properties required for the simulations are runoff curve number for average antecedent moisture conditions (CN2) and the organic carbon content of the grass and thatch. Both of these parameters depend on grass height and thatch thickness, which were assumed to be 11 and 8 mm, respectively, as in Haith and Rossi *(7)*. Using the procedures given in Haith *(1)*, these values produce a curve number of 67 and organic carbon content of 10,200 kg/ha. The curve number selection also assumes a hydrologic group C (relatively poor drained) soil.

Pesticide Characteristics

The two pesticide properties required by TurfPQ, bio-degradation half-life and partition coefficient, are relatively easily obtained. The partition coefficient is computed from turf organic carbon content and K_{oc}, the organic carbon partition coefficient. Half-lives and K_{oc} values are available from general databases *(8, 9, 10)*. Application amounts and timing are also required for the simulations, and these can be quite difficult to obtain. Although application rates are specified by labels *(11)*, a wide range is often given, corresponding to use against different pests. Because it is likely that the chemicals will typically be used against a variety of pests, the median or mid-range label value, converted to g/ha of active ingredients, was used in the simulations.

Timing, or frequency of applications, is less straightforward. Publicly available application records are very rare, and we know of no general databases. In the absence of other information, we based simulation applications on label suggestions of prophylactic applications at regular intervals to control multiple pests. These applications will almost certainly be more frequent than those used by many turf managers, particularly those following integrated pest management programs. The major determinants were pesticide type (herbicide, fungicide, insecticide), growing season, as shown in Table I, and application intervals and annual or seasonal limits specified by the labels. Generally, longer growing seasons result in more applications of a pesticide, unless limited by label.

Herbicides are divided into pre-emergent and post-emergent. The former is applied as a single application on the first day of the growing season. Post-emergent herbicides such as MCPP are assumed to be applied in the middle of each of the first two months of the growing season and once in the last or next to last month of growing season if allowed by the label.

Fungicide applications were based on preventative control of diseases such as dollar spot, summer patch, brown patch, and leaf spot. Applications were generally started in the middle of the second growing month, and if permitted by label, continued every 15 days through the middle of the next to last growing month. Otherwise, label limits applied, as was the case with fenarimol, which was applied every 30 days.

As with fungicides, repetitive preventive applications are assumed for insecticides, which are used to control a range of pests (grubs, chinch bugs, cutworms, webworms, billbugs) which occur mainly in late Spring and Summer. For insecticides such as carbaryl, this suggests a mid-month application starting in the second growing season month and continuing through September. Pesticide properties, rates and application frequencies for the four simulated chemicals are given in Tables II and III.

Table II. Pesticide Properties and Application Rates

Pesticide	*Rate per Application (g/ha)*	*Half-Life (d)*	*K_{oc} (cm^3/g)*
MCPP	860	10	20
Fenarimol	760	840	760
Iprodione	4580	50	670
Carbaryl	8000	17	290

Application rates and frequencies differ markedly for these four chemicals. For example, the total annual pesticide application for the Atlanta site ranges from 2580 g/ha for MCPP to 40,000 g/ha for carbaryl. Application frequency is lowest for Madison because of its short growing season. This produces much lower inputs of the 2 fungicides than seen at the other sites. The large number of fungicide applications for Fresno may seem inconsistent with its dry climate, which would not typically favor plant diseases. However, the regular irrigation inputs needed to maintain Fresno fairways produce the warm, humid conditions required for disease development.

The pesticides differ markedly in their persistence and adsorption characteristics (half-lives and K_{oc}). MCPP is an ephemeral chemical that is only weakly adsorbed, and unlikely to remain long in the turf. Carbaryl is similarly short-lived, but more strongly adsorbed and thus less readily leached. Both fungicides are relatively strongly adsorbed, and fenarimol is very long-lived.

Table III. Pesticide Application Frequency

	Number of Applications			
Pesticide	*Atlanta*	*Fresno*	*Madison*	*Olympia*
MCPP	3	3	3	3
Fenarimol	5	7	3	4
Iprodione	6	6	3	5
Carbaryl	5	6	4	4

Organization of Results

Each simulation experiment produces 100 years of daily estimates of water volumes and pesticide mass loads in fairway runoff. The information was summarized by annual and monthly means and by the annual maximum daily load (AMDL) of pesticide runoff. The AMDL is the largest one-day runoff load produced in a year. The 100 values of AMDLs are then used to assign return periods to these extreme event. Thus the 1 in 10 year AMDL would be expected to be exceeded on the average of once in 10 years, or 10 times in 100 years.

Simulation Results

Annual Water Balances

Mean annual water inputs and runoff from the 100-year simulations are given in Table IV. Overall, regional differences in weather and hydrology for these sites are rather substantial. Runoff was minimal for Fresno because most water input was from the regular addition of moderate irrigation amounts rather than large precipitation events. Runoff was 3-4% of total water inputs at the other sites, and 40-50% of the runoff occurred during the growing seasons at Atlanta and Madison. Although Olympia had significant annual runoff, very little occurred during the growing season, when pesticides were being applied.

Table IV. Mean Annual Fairways Water Inputs and Runoff

Location	*Precipitation*	*Irrigation*	*Total*	*Year Runoff*	*Growing Season Runoff*
	mm				
Atlanta	1281	435	1716	77	34
Fresno	272	771	1043	2	<1
Madison	789	307	1096	32	16
Olympia	1304	330	1634	65	4

Annual Pesticide Runoff

The mean annual pesticide mass loads in runoff from the 100-year simulations are shown in Figure 1. To a considerable extent, these results reflect the differences in application and runoff water amounts. Chemicals such as

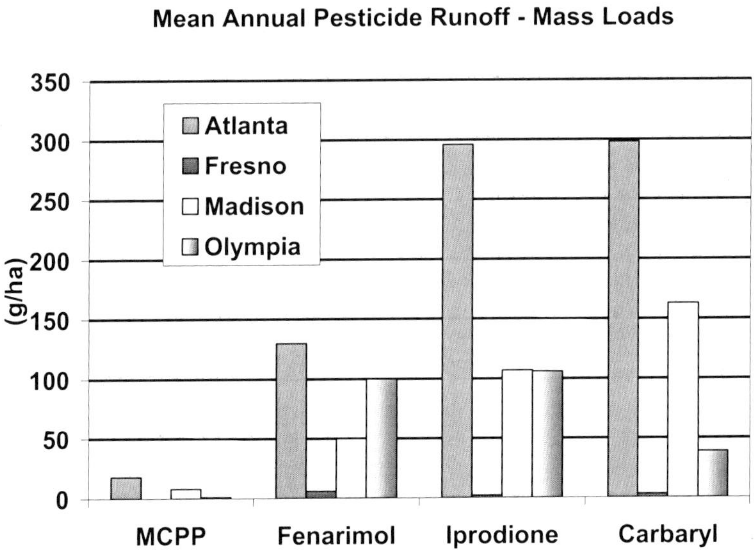

Figure 1. Mean annual runoff of four pesticides at four fairways sites.

iprodione and carbaryl, which are applied in greatest amount, are most likely to be seen in runoff. Similarly, sites such as Fresno, which have very little runoff, have correspondingly small amounts of pesticide loss. However, regional differences are not always clear-cut. Madison and Olympia produced almost equal amounts of iprodione runoff, but differed greatly in fenarimol and carbaryl runoff. Fenarimol runoff was 50% higher in Olympia but carbaryl runoff was 4 times higher in Madison. These apparently inconsistent results reflect the chemicals' properties and occurrences of runoff water. Carbaryl is short-lived and most likely to be lost during the growing season in which it is applied.

Growing season runoff is much higher in Madison, so there will be greater opportunities for loss. Conversely, the persistence and strong adsorption of fenarimol means that it is likely to remain available for runoff following the growing season, when Olympia experiences much greater runoff than Madison.

Comparisons of mean annual pesticide runoff look rather different when the mass loads are normalized with respect to the total annual application, as shown in Figure 2. Three of the pesticides, MCPP, iprodione and carbaryl, show similar tendencies for loss in runoff, but these losses are much smaller than those seen

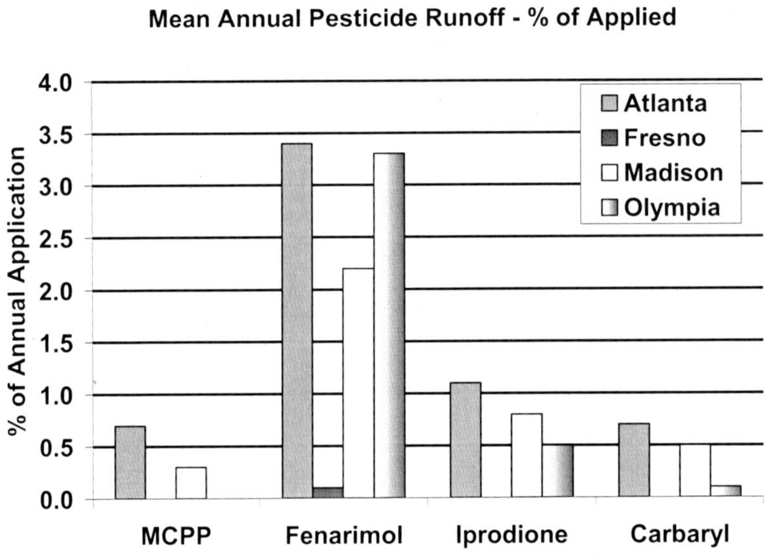

Figure 2. Mean annual fairway runoff pesticides expressed as a percentage of annual application.

for fenarimol. It is apparent that the relatively larger mass runoff loads of iprodione and carbaryl that were seen in Figure 1 were more a result of the larger applications of the chemicals than their inherent propensities for loss.

The regional differences in the normalized runoff results are consistent with those observed for mass loads. Atlanta still produces the greatest runoff losses, but the differences with other locations are less striking. Atlanta mass loads were more than twice as large as Madison's for all chemicals, but when expressed as a percentage of annual applications, the differences are much less, particularly for iprodione and carbaryl. Averaged over all chemicals, mean annual losses by location are Atlanta – 1.5%, Fresno – <0.1%, Madison and Olympia – 1.0%. Mean chemical losses, averaged over location, are MCPP – 0.3%, fenarimol – 2.3%, iprodione – 0.6% and carbaryl – 0.3%.

Pesticide Runoff in Extreme Events

Although water quality impacts are often measured in terms of mean annual loads such as those shown in Figure 1, these indicators may be of limited value for turf pesticide runoff. Most turf systems, including the fairways modeled in these simulation experiments, produce water runoff infrequently, and significant pesticide runoff is produced only when one of these events coincides with a high level of available chemical in the turf foliage and thatch. There are typically few such occurrences in any year, but it is these short-term phenomena that are responsible for any impact that pesticide runoff will have on surface receiving waters. Mean annual loads are useful indicators of the relative likelihood of pesticide runoff, but they are imperfect measures of impact.

Figure 3 shows the 1 in 10 year, 1-day pesticide runoff event modeled for the four chemicals and sites. This is the event that is likely to occur, on average, once every 10 years. These results look very different than the mean annual values shown in Figure 2. Fenarimol remains the chemical with highest percentage loss at three locations, but MCPP runoff in Atlanta exceeds that of the other chemicals. Atlanta no longer sees the largest losses for all pesticides. Madison produces the greatest percentage losses of fenarimol and iprodione, even though runoff water volume is much lower than for Atlanta and Olympia.

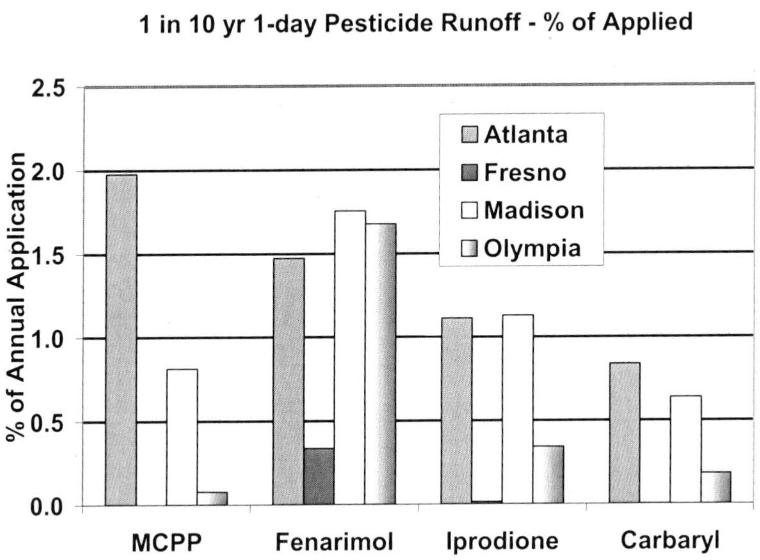

Figure 3. One in ten year pesticide runoff events.

The regional and chemical variations seen in the extreme event results in Figure 3 are not readily explained by differences in pesticide properties or mean weather conditions. Rather, they are influenced by the nature of the precipitation and snowmelt events at the site, and these in turn are determined by the probabilistic structure of the weather. Are large storms more likely to occur shortly after pesticide application? Is a site characterized more by large numbers of small storms rather than by rare large events? The effects of these weather characteristics on pesticide runoff and the resulting water quality impacts are essentially unpredictable, but we can observe them through long-term simulations such as these.

Conclusions

Although field and simulation experiments have comparable objectives and designs there is at least one profound difference. Field experiments measure reality and simulation experiments estimate reality. Granted, measurements can be error-prone and misconstrued, and simulation models are often tested through field experiments, but still, one is real and the other is not.

So what are the conclusions that can be drawn from these simulation experiments? First, it would be foolhardy to greatly trust the absolute values of the results. Because of the intensive pesticide application frequencies and poorly drained soils assumed in the simulations, the resulting pesticide runoff loads are likely to be larger than those seen on many sites. Further, although the TurfPQ model is relatively accurate on average, its estimates can differ substantially from field measurements for any one site or chemical. For example, we know that its estimates of runoff of strongly adsorbed pesticides are often too high *(1,2)*. However, the model does better at estimating differences among chemicals and sites, and that is the critical attribute for the present study.

The major conclusion of the simulation experiment is that substantial differences exist in turf pesticide runoff among sites and chemicals. Stated another way, conclusions about runoff of a pesticide at one location cannot safely be extrapolated to another pesticide or location. This is true whether we are talking of mean mass loads, percentage losses or extreme events. Some of the differences can be explained by chemical properties and annual weather and hydrology conditions, but others, including those seen in extreme pesticide runoff events, are unpredictable, and only become apparent with long-term observations.

This is not a happy conclusion for the turf manager or government regulator wishing to adopt general guidelines for environmentally safe pesticide use. It argues that each combination of chemical, application regime and site is unique. Field experiments for each of these situations would be immensely impractical, but simulation experiments are much less forbidding. The simulation protocol described in this paper is a straightforward intuitive process that is accessible to anyone with computer skills and basic knowledge of pesticides and turf. The process could be made even more accessible if captured in a web-based interactive system. It may be time to admit that generalizations regarding environmental impacts of turf pesticides are neither desirable nor necessary. The methods and data are available for quantitative analyses of each unique situation.

References

1. Haith, D. A. *J. Environ. Qual.* **2001,** *30,* 1033-1039.
2. Haith, D. A. *J. Environ. Qual.* **2002,** *31*, 701-701.
3. Hanson, C. L.; Cumming, K. A.; Woolhiser; D. A., Richardson; C. W. *Microcomputer Program for Daily Weather Simulation in the Contiguous United States;* ARS-114. UDSA-ARS: Washington, D.C. 1994.
4. National Oceanographic and Atmospheric Administration. *Comparative Climatic Data.* URL http://ols.nndc.noaa.gov/plolstore/. 2005.
5. Koss, W. K., Owenby, J. R., Steurer, P. M., Ezell, D. S. *Freeze/Frost Data;* Climatology of the U.S. No. 20, Supplement No. 1. National Climatic Data Center: Ashville, NC. 1988.
6. Hamon, W. R.. *Proc. Am. Soc. Civ. Eng., J. Hyd. Div.* **1961,** 87(HY3), 107-120.
7. Haith, D. A.; Rossi, F.S. *J. Environ. Qual.* **2003,** *32,* 447-455.
8. Agricultural Research Service. *The ARS pesticide properties database.* URL http://www.arsusda.gov/acsl/services/ppdb/. 2005.
9. *The Pesticide Manual;* Tomlin, C., Ed. British Crop Protection Council: Farnham, U.K. 2000.
10. Natural Resources Conservation Service. *Pesticide properties database.* URL http://www.wcc.nrcs.usda.gov/pestmgt/. 2005
11. Vance Communication. *Greenbook Database of Labels, MSDS, and Supplemental Labels,* URL http://www.greenbook.net/. 2005

Chapter 13

Development and Testing of a Comprehensive Model of Pesticide Losses from Turf

Antoni Magri[1], Douglas A. Haith[1], A. Martin Petrovic[2], Laosheng Wu[3] and Robert L. Green[4].

[1]Department of Biological and Environmental Engineering, Cornell University, Ithaca, NY 14853
[2] Department of Horticulture, Cornell University, Ithaca, NY 14853
[3]Department of Environmental Sciences, University of California-Riverside, Riverside, CA 92521
[4]Department of Botany and Plant Sciences, University of California-Riverside, Riverside, CA 92521

Development of the TURFP simulation model addresses the need for an engineering tool that can simultaneously evaluate pesticide losses via runoff, leaching and volatilization from turf, as a basis for risk assessment and water quality management. TURFP integrates previously tested and published pesticide runoff and volatilization models with a new pesticide leaching component. The leaching model uses a simple approach based on mass balances, and requires soil and pesticide data that are readily available in published databases. Default values are suggested so that the model may be run in situations where measured data is not available for calibration. The uncalibrated leaching model was tested using default parameter values for 427 drainage and 469 pesticide leachate measurements taken at two sites, involving six pesticides and four soils. Mean predicted pesticide loss via leaching was 0.28% of the applied pesticide amounts, compared to an observed mean of 0.25%. The leaching component captured the dynamics of drainage and pesticide leaching occurrences reasonably well, with R^2 values of 0.69 for drainage and 0.47 for pesticide leaching. Strengths of the leaching component are that it requires few input parameters and appears to predict pesticide leaching adequately without site-specific calibration.

Introduction

Turf is frequently recognized as one of the most intensively managed biotic systems (*1, 2*), and can account for 80% of the pervious surfaces in urban areas, of which approximately half may be subjected to high input management (*3*). Additionally, pesticide application rates may be 3 to 8 times higher on golf courses, and 3 times higher on home lawns, than those applied to agricultural land (*4, 5*). Pesticide concentrations in water from urban areas and golf courses have been found to exceed environmental and drinking water standards (*6, 7*). Numerous field studies indicate that up to 25% of the applied pesticide may leave the turf system via runoff or leachate after heavy rain or irrigation events (*2, 8, 9, 10, 11, 12, 13, 14, 15, 16, 17, 18, 19, 20*). Leaching from turf over many hydrologic events can account for up to 62% of the applied pesticide (*21*).

While monitoring and field studies provide valuable data, they are time consuming and expensive to implement, and their results are not easily extrapolable to different geographic, climatic, soil and management conditions. An alternative approach is to develop and use mathematical models to predict the fate and transport of pesticides. Models applied to turf systems include Erosion-Productivity Impact Calculator (EPIC), Groundwater Loading Effects of Agricultural Management Systems (GLEAMS), Pesticide Root Zone Model (PRZM), Leaching Estimation and Chemistry Model (LEACHM), OPUS and the Soil and Water Assessment Tool (SWAT). None of these were developed specifically to simulate turf systems, and their application has involved adapting them to include thatch, usually by treating it as an additional layer of soil. This is inconsistent with a fundamental reality of turf systems – a dominant effect of plant material, rather than soil, on chemical fate and transport.

Results obtained using some of these models to simulate turf systems were as follows: GLEAMS and PRZM predictions of pesticide concentrations in soil water varied between factors of 2 or 3 to several orders of magnitude for field measurements in Georgia (*22*, *23*). LEACHM's predictions in Ontario were similar to observed concentrations, but some instances were underpredicted by factors of 2 to 3 (*24*). Pesticide leaching predictions by KTURF (artificial neural network model) were within about 4% of the testing case values, but model inputs to the model do not include soil characteristics (*25*) and it is unclear how well the model would perform at other sites without further user training. A two-site non-equilibrium model (*26, 27*) achieved R^2 values between 0.7 and 0.94, but test data were limited and it is difficult to generalize these results.

TurfPQ is a validated model that simulates pesticide losses in runoff (*28*). In the model, the Unites States Department of Agriculture Soil Conservation Services, (Renamed the Natural Resources Conservation Service) Curve Number runoff estimation method uses parameter values developed to specifically reflect the effects of turf vegetation and thatch on runoff volumes (*29*). TurfPQ's uncalibrated performance was assessed for six pesticides applied at five study sites located in four states, and achieved an R^2 of 0.65 (*28*). A validated model for simulating pesticide volatilization from turf has also been developed. The pesticide volatilization model's performance was tested for eight pesticides and achieved an R^2 of 0.67 (*30*).

There is currently a need for simulation models that can be easily applied to turf scenarios without requiring extensive parameterization and calibration, as a basis for risk assessment, total maximum daily load and turf management studies. This paper describes the development and testing of a leaching component that has been coupled with TurfPQ and a volatilization model (*31*) in order to generate a comprehensive model of pesticide loss from turf. The leaching component is based on a simple approach that requires few input parameters. Calibration is optional and default parameter values are provided, so that the model may be applied where data are unavailable for calibration. The following sections give an overview of the leaching model, the plot studies used to test its performance, and comparisons between model predictions and field observations.

Methods

Leaching Model

The pesticide leaching model is comprised of two main parts: a soil water balance model and a soil chemistry model. The soil water balance model considers infiltration from the turf and thatch layer, evapotranspiration and drainage. The soil chemistry model simulates partitioning into adsorbed and dissolved phases, microbial decay and movement through the soil medium. Both submodels are coupled so that pesticide movement is dependent on water movement through the soil. For computational purposes, the model splits the soil into 1 cm thick layers and uses a daily timestep for all calculations. The following sections provide a brief overview of the leaching model's components.

Soil Water Balance Model

TURFP uses a mass balance approach that considers three major processes: infiltration, evapotranspiration and drainage,

$$\theta_{i,t+1} = \theta_{i,t} + PC_{i-1,t} - ET_{i,t} - PC_{i,t} \qquad (1)$$

where $\theta_{i,t}$, $PC_{i,t}$ and $ET_{i,t}$ are the soil water content, drainage and evapotranspiration, respectively, of layer i on day t (cm). When precipitation, irrigation and/or snowmelt events occur, the equations developed for TurfPQ (*28*) partition water into runoff and into infiltration, which enters the top layer of soil directly.

ET is estimated on a daily basis using the Hargreaves-Samani equation (*32*). This approach was found to overpredict ET with respect to the American Society of Civil Engineers (ASCE) standardized estimate of ET for many locations of the United States (*33*). Correction factors (based on reference *33*) may be used to adjust local ET estimates. Monthly crop coefficients (K_c) are used in TURFP to account for differences in evapotranspiration between warm and cool season turfgrasses.

Evapotranspiration from each layer of soil is limited by the root mass that the layer contains. This is modeled by assuming that root mass decreases exponentially with depth (*34*). Exponential decrease in water uptake with depth under turf has been observed in field and laboratory experiments (*35*). Thus, water loss via evapotranspiration is heavily weighted towards the top layers of soil in TURFP.

Evapotranspiration is also limited by the amount of water available in each layer of soil. Water extraction is decreased linearly if soil moisture falls below the midpoint between field capacity and wilting point. This approach is similar to that used in other models (e.g. PRZM3 and EPIC). Once the water content reaches wilting point, no further water can be extracted from the soil layer.

If water content in a soil layer still exceeds field capacity after evapotranspiration has occurred, excess water (above field capacity) is allowed to drain to the layer below, using a tipping bucket approach.

The sequence of events is arbitrarily set to: water enters a layer of soil, evapotranspiration occurs, and excess water is drained to the layer below. This is repeated until the bottom layer of the soil profile is reached. TURFP operates under the assumption that the entire soil profile drains in a 24 hour period, which is supported by findings for rooting media 30 to 40 cm deep (*36*).

Soil Chemistry Model

The leaching component of TURFP simulates adsorption and decay of pesticides, and couples their movement to the flow of water through the soil. The mass balance equation is:

$$P_{i,t+1} = [P_{i,t} + PL_{i-1,t} - PL_{i,t}]\,[1 - \exp(-\alpha)] \tag{2}$$

where $P_{i,t}$ is the pesticide mass in layer i at the beginning of day t, $PL_{i,t}$ is the pesticide mass leached from layer i during day t and α is the pesticide decay rate (d^{-1}). Pesticide entering a layer of soil is added to the pesticide previously existing in the layer. The total pesticide mass is then mixed homogeneously throughout the layer and partitioned into adsorbed and dissolved phases using a linear instantaneous equilibrium approach, dependent on the organic carbon partition coefficient (K_{oc}) of the pesticide. The amount of water available for the pesticide to dissolve in is the water remaining after ET has occurred, but before excess water has drainage to the layer below. When excess water drains, it carries a proportion of the dissolved pesticide with it.

Pesticide remaining in layer i then undergoes first order decay. The pesticide's decay rate $\alpha = 0.693/\tau_{1/2}$, where $\tau_{1/2}$ is the pesticide's half-life (d). In order to include the effects of decreased microbial activity with depth, values of the decay rate α are reduced exponentially from the surface soil layer downwards, following (*37*). The decay rate is set to the pesticide's microbial decay rate at the soil surface, and decreases to the pesticide's hydrolysis rate at a depth of 1 m. Hydrolysis is assumed to be the major mechanism of pesticide decay below a depth of 1 m, which is similar to the approach used in the Root Zone Water Quality Model (RZWQM, *38*).

Water and pesticides leaving the bottom layer in the simulated soil profile are assumed to enter a shallow aquifer, and can thus affect groundwater and, possibly, surface water quality via groundwater discharge to streams.

Default Parameter Values

Soil input parameters are bulk density (BD, g/cm), organic matter (OM, % by weight), field capacity (FC, cm/cm) and wilting point (WP, cm/cm), and must be provided for each of the soil horizons included in the simulated soil profile. Soil horizons are split into 1 cm thick computational layers, each of which receives parameter values corresponding to the soil horizon that it belongs to. Characteristics of native soil horizons are available from soil surveys, or can be obtained by using programs such as Soil Water Characteristics (*39*), which estimates parameters such as bulk density, field capacity and wilting point based on texture, organic matter and other variables (*40*). Values for artificial rooting media can be estimated based on data for sand-peat mixtures (*41*).

K_c values for turf are scarce, but some have been published (*42, 43*), and can be used directly with ET estimates for turf (*33*). K_c values developed by The Irrigation Association (*42*) are suggested as default values.

The root mass model described in (*34*) requires one parameter: the soil depth above which 63% of the total root mass is located. Based on studies for bermuda and zoysiagrass (*44*), for Kentucky bluegrass (*35*) and on values recommended for modeling well-established turf (*45*), the parameter value was found to vary between 5 and 10 cm. Eight cm provided a good fit to the reported data, and is suggested as a default value.

Other input parameters for TURFP appearing in Equations (1) and (2) and related explanations are the same as those for TurfPQ (*28*) and the pesticide volatilization model (*31*). They include the pesticide half-life (d), hydrolysis rate (d^{-1}), K_{oc} value (cm^3/g), molecular weight, vapor pressure (kPa), and the state of the pesticide as it was applied to the turf (solid or liquid). Values can be obtained for many pesticides from the USDA Pesticide Properties Database (*46*) or compendia (*47, 48, 49*).

Pesticide runoff estimation also requires values of the curve number for average moisture conditions (CN2, based on soil hydrologic groups), and of the monthly organic carbon content of the turf vegetation and thatch layer, for which (*28*) provides default values and estimation procedures. Finally, an indication of growing and dormant seasons must be provided, and can be determined using historical first freeze and last frost dates.

Field Studies

Drainage volumes, pesticide leaching data and the necessary parameters to run the TURFP model were available for two plot studies. The following provides a brief description of each.

In Ithaca NY, the herbicide mecoprop (2-(4-chloro-2-methyl-phenoxy) propionic acid) was applied to creeping bentgrass (*Agrostis palustris* Huds.)

growing on three different types of soils: Hudson silt loam (fine illitic, mesic Glossaquic Hapludalf), Arkport sandy loam (coarse-loamy, mixed, active, mesic Lamellic Hapludalf) and sand. Vegetation was maintained at a fairway height of 12 mm, and thatch was nonexistent during the trial (A. M. Petrovic, personal communication 2005). Mecoprop was sprayed at a rate of 3107 g a.i./ha on the 24th of September 1991. An automatic rainout shelter was used to exclude natural rainfall, and irrigation schemes simulated historically moderate and high precipitation patterns. Treatments were replicated four times, and monitoring continued until March 1992 (*21*).

In Riverside CA, creeping bentgrass lysimeters were sprayed with the fungicides metalaxyl (methyl N-(methoxyacetyl)-N-(2,6-xylyl)-DL-alaninate) and chlorothalonil (2,4,5,6-tetrachloro-1,3-benzenedicarbonitrile) in 1995, and with the insecticides trichlorfon (dimethyl-2,2,2-trichloro-1-hydroxy-ethylphosphonate) and chlorpyrifos (O,O-diethyl O-3,5,6-trichloro-2-pyridylphosphorothioate) in 1996 and 1997. Rooting media consisted of 45 cm of a 9:1 sand-peat mixture, 43 cm of pea gravel and 7 cm of gravel. Thatch and mat were 2 and 3 cm thick, respectively. Organic carbon contents were 6.1, 3.08 and 0.08% by mass in 1995, and 6.6, 3.4 and 0.3% in 1996, for the thatch, mat and soil layers, respectively. Pesticide application rates were 1531 g a.i./ha of metalaxyl and 12740 g a.i./ha of chlorothalonil, and were applied on September 27 1995. 7650 g a.i./ha of trichlorfon and 1530 g a.i./ha of chlorpyrifos were applied on June 4 1996 and on July 9 1997. Vegetation was maintained at 5 mm height and irrigated to prevent visual drought symptoms. Drainage volumes were collected daily and combined (over 1 to 3 days) for a period of 146 days in 1995, 73 days in 1996 and 79 days in 1997. Four replications of the experiment were performed in 1995 and 1996, and three in 1997 (*2, 20*).

Model Testing

As mentioned, Hargreaves-Samani ET exceeded standardized ASCE ET estimates by 25% for Ithaca, NY and by 12% for Santa Maria, CA during the growing season (*33*). Overpredictions were corrected in all model simulations by multiplying ET by 0.8 for Ithaca, NY and 0.89 for Riverside, CA. K_c values (0.61, 0.69, 0.77, 0.84, 0.9, 0.93, 0.93, 0.89, 0.83, 0.75, 0.67 and 0.59 for January to December, respectively) corresponding to cool season turf (*42*) were used for all simulations.

Appropriate CN2 values were determined based on the hydrologic group of each soil (group A for the sand and sand-peat mixture, group B for Arkport and group C for Hudson soils), and the height of the vegetation at each site. Thatched short grass was used for the California site, and short grass with > 75% ground cover was used for the New York site. Resulting CN2 values were 35 for the sand-peat mixture used in CA, 39 for the sand used in NY, 61 for the Arkport sandy loam and 74 for the Hudson silt loam (*28*).

As monthly estimates for turf vegetation and thatch organic carbon (OC) were not available, one value was determined and used for all months in each simulation. 1308 kg OC/ha was used for the New York site, based on the vegetation height default method (*28*). Values for the California site were 9695 kg OC/ha in 1995 and 10445 kg OC/ha in 1996 and 1997, based on measured % OC of the thatch layer and an assumed bulk density of 0.75 g/cm^3 for thatch (*50*). The mat layer in CA was modeled as a 3 cm layer of soil, with 5.3% OM in 1995 and 5.9% OM in 1996 and 1997, based on site measurements of % OC (*2, 20*).

Pesticide parameter values are given in Table I. Values of degradation half-lives and adsorption coefficients determined for the CA site (*2, 20*) were not used for model testing, as site specific values such as these are not likely to be available at other sites where the model may be applied. Soil data is summarized in Table II. Growing seasons were determined to be from May to September for the NY site, and from March to November for the CA site, based on analyses of

Table I. Pesticide Parameter Values

Pesticide	*Half-life (days)*	*Hydrolysis Rate (day^{-1})*	*K_{oc} (cm^3/g)*	*Molecular Weight*	*Vapor Pressure (kPa, 25°C)*
Mecoprop	10.0†	0 (stable) †	18.5†	214.60	3.1E-7
Metalaxyl	40.0	0 (stable)	171.0	279.34	7.5E-7
Chlorothalonil	20.5†	0 (stable)	5000.0	265.92	7.6E-8
Trichlorfon	6.4	0.489	15.0	257.44	2.7E-7
Chlorpyrifos	30.5	0.024	9930.0	350.62	2.5E-6

SOURCES: † (*47*), all other values from (*46*).

Table II. Soil Properties

Site	*Layer*	*Depth†* (m)	*OM†* (%)	*BD‡* (g/cm^3)	*FC‡* (m/m)	*WP‡* (m/m)
NY	Sand	0.37	0.8	1.71	0.126	0.056
NY	Arkport sandy loam	0.37	4.4	1.26	0.299	0.124
NY	Hudson silt loam	0.37	5.8	1.14	0.366	0.144
CA	Mat	0.03	5.3/5.9[1]	0.90	0.259	0.097
	Sand-peat mix	0.45	0.14/0.5[1]	1.45§	0.140§	0.060§
	Pea gravel	0.43	0.0	1.60*	0.030*	0.030*
	Gravel	0.07	0.0	1.60*	0.030*	0.030*

NOTE: [1] The first value is for 1995, second value is for 1996 and 1997.

SOURCES: † (*2, 20, 21*). ‡ (*39*). § Estimated based on (*41*). * Assumed based on properties for coarse sand. FC set equal to WP as the particles are too coarse to contribute to the water holding capacity of the soil profile.

historical weather data. Daily irrigation data were provided by Petrovic and Wu (personal communications, 2005), and daily minimum and maximum air temperatures were obtained from the closest weather station to each site (Cornell

University for the New York experiment, and Riverside for the California experiment).

Results and Discussion

Predicted values were based on blind simulations using the default parameter values described previously. As such, the results give an indication of how the model might be expected to perform when site data is limited. No attempts were made to calibrate the model, and it is likely that model performance would improve when site measured data is used and the parameters are calibrated to match observed leaching.

Drainage Comparisons

The 427 observed and simulated drainage event values are summarized in Table III by showing the average observed and model predicted drainage for each treatment. The model underpredicted drainage in all but one experiment. The nonparametric Mann-Whitney test (*51*) indicated that the observed and predicted means (for the 427 events) were significantly different (α=0.05). It should be noted that the magnitudes of the mean values were considerably larger for the Arkport and Hudson soil treatments because the time intervals over which drainage volumes were accumulated before collection were longer than for the other experiments.

Graphic analysis of the observed and predicted drainage showed that points generally fell close to the 1:1 line, but that the larger events tended to be underpredicted by the model. These correspond to the Arkport and Hudson soils under the high precipitation treatment, and may indicate errors in parameterization of soil properties for those two soils.

Table III. Summary of Observed and Simulated Drainage

		Mean drainage	
Site	*Experiment*	*Observed (mm)*	*Simulated (mm)*
NY	Sand high prcp.	3.1	2.3
	Sand moderate prcp.	2.0	1.7
	Arkport high prcp.	13.5	11.7
	Arkport moderate prcp.	11.1	10.6
	Hudson high prcp.	13.0	11.1
	Hudson moderate prcp.	13.0	10.2
CA	1995	7.7	7.7
	1996	6.0	3.7
	1997	4.8	3.3
All	Mean	5.5	4.5
Events	Standard deviation	6.9	6.6
	R^2		0.69

Pesticide Leaching Comparisons

The 469 observed and simulated pesticide leaching values are summarized in Table IV by showing the average observed and model predicted percentage of the applied pesticide mass leached from the bottom of the soil profile for each treatment. The Mann-Whitney test indicated that observed and predicted means (for the 469 events) were not significantly different (α=0.05), but the results showed instances of considerable overprediction (mecoprop applied to the Arkport and Hudson soils in NY, and metalaxyl applied to a sand-peat mixture in CA). Overprediction of metalaxyl leaching was likely due to the very low organic matter content of the sand-peat mixture in 1995 (0.14%), versus higher values observed in 1996 (0.5%) and a default estimated value of 0.72% for sand-peat mixtures of 10% peat by volume. Simulations using 0.5 and 0.72% OM showed that metalaxyl leaching is very sensitive to this parameter, probably due to its low K_{oc} value, which caused it to be poorly retained in the thatch layer and to leach readily into the soil. Chlorothalonil, also applied in 1995, did not prove to be sensitive to changes in soil % OM, most likely because it was strongly retained in the thatch layer due to its higher K_{oc} value.

It is possible that overpredictions for the Arkport and Hudson soils were caused by uncertainty in the soil OM contents as well, although simulations were not performed to investigate these discrepancies. A second possibility is that the K_{oc} value for mecoprop (MCPP) might be inappropriate for the NY site, and that the pesticide would be more strongly retained in soils with higher OC contents (Arkport and Hudson).

Table IV. Summary of Observed and Simulated Pesticide Leaching

Site	Experiment	*Pesticide Leached* *Observed (% app)*	*Simulated (% app)*
NY	MCPP Sand high prcp.	2.80	2.46
	MCPP Sand mod. prcp.	2.81	2.41
	MCPP Arkport high prcp.	0.02	0.43
	MCPP Arkport mod. prcp.	0.05	0.31
	MCPP Hudson high prcp.	0.06	0.16
	MCPP Hudson mod. prcp.	0.04	0.00
CA	Metalaxyl 1995	0.01	0.36
	Chlorothalonil 1995	0.00	0.00
	Trichlorfon 1996	0.00	0.00
	Chlorpyrifos 1996	0.00	0.00
	Trichlorfon 1997	0.00	0.00
	Chlorpyrifos 1997	0.00	0.00
All	Mean	0.25	0.28
Events	Standard deviation	1.22	1.27
	R^2		0.47

Graphic analysis of the observed and simulated percentages of pesticide mass that leached showed considerable scatter around the 1:1 line. Considering that the observed and predicted means are similar, the scatter may indicate that the timings of the observed and simulated events did not correspond well. It is possible that the assumption that the entire soil profile drains in 24 hours is not accurate for soils deeper than 0.3 to 0.4 m. Delayed exit of the pesticide from the soil profile, possibly due to strong adsorption to thatch and excessively reduced decay rates in the soil, may explain a portion of the scatter.

Overall model performance was also evaluated using the Nash-Sutcliffe efficiency measure (*52*), where observed and predicted are values for individual events (drainage or percentage of applied pesticide mass that leached), and observed is the mean of the observed values.

$$E = 1 - \frac{\sum(\text{predicted} - \text{observed})^2}{\sum(\text{observed} - \overline{\text{observed}})^2} \qquad [3]$$

Models are efficient predictors when E approaches 1, and are no better than using the mean of the observations as a predictor when $E = 0$. The leaching model efficiently predicts drainage ($E = 0.72$), but is less efficient at predicting pesticide leaching ($E = 0.34$). The efficiency of pesticide leaching prediction increased to 0.4 when the % OM for the sand-peat mixture was increased from 0.14 to 0.5% for the 1995 experiments in CA, affecting the strongly overpredicted metalaxyl leaching (as discussed previously).

Sources of Error

Possible sources of error include uncertainties due to the large variability in pesticide parameter values (*48*). As noted for metalaxyl, leaching of pesticides not strongly retained in the turf vegetation and thatch layers appears to be sensitive soil OM. The model's assumption that the soil profile drains in 24 hours may also lead to errors in drainage and pesticide leaching. Finally, the exponential effect of depth on the decay rates of the pesticides may decrease pesticide decay rates too quickly, and it is possible that a linear approach may better describe the decrease in decay rates for soils under turf.

Future Development of TURFP

Further work will include refinement and testing of the leaching component with other datasets. Possible refinements include the use of travel time to delay flow through the soil profile, and adjustments to the depth effect on pesticide decay rates. The runoff and volatilization routines will be retested in order to verify that they perform adequately as parts of a comprehensive model. Further model testing will be performed using site-measured data and calibration in order to assess model performance in scenarios where detailed site data and measurements of pesticide loss exist.

Conclusions

A pesticide leaching component was developed and integrated with tested pesticide runoff and volatilization models to create the TURFP simulation model. The leaching component requires few parameters (soil BD, %OM, FC, WP, warm and cool season turf K_c values, and a parameter describing the root distribution). Default values are provided for K_c and the turf root distribution parameter, while other parameter values may be obtained from soil surveys.

Tests using default values to simulate 427 drainage and 469 pesticide leachate observations collected at two different sites demonstrated that the model underpredicted drainage and slightly overpredicted pesticide leaching. Inaccuracies in predicting pesticide leaching may be due to the model's sensitivity to pesticide and soil parameter values. Dynamics were captured reasonably well, with $R^2 = 0.69$ and a model efficiency of 0.72 for drainage, and $R^2 = 0.47$ and an efficiency of 0.34 for pesticide leaching.

TURFP allows a complete description of pesticide fate and transport in turf systems. Its other strengths include that it is based on components which have been developed specifically for turf systems and which perform well when tested against field data, without requiring calibration. Additionally, default parameter values are available for all components of the model, so that it may be run in situations where site specific information for calibration is unavailable, providing a basis for risk assessment and water quality management studies.

References

1. Gardner, D.S.; Branham, B.E.; Lickfeldt, D.W. *Crop Sci.* **2000**, 40, 1333-1339.
2. Wu, L.; Green, R.L.; Liu, G.; Yates, M.V.; Pacheco, P.; Gan, J.; Yates, S.R. *J. Environ. Qual.* **2002**, 31, 889-895.
3. Schueler, T. R. In *The Practice of Watershed Protection;* Schueler, T. R.; Holland, H. K., Ed. Center for Watershed Protection, Ellicott City, MD. 2000, pp. 649-654.
4. Koppell, G.O. New York State Department of Law, New York. 1994. URL http://www.oag.state.ny.us/environment/golf95.html
5. Schueler, T. R. In *The Practice of Watershed Protection;* Schueler, T. R.; Holland, H. K., Ed. Center for Watershed Protection, Ellicott City, MD. 2000, pp. 673-675.
6. U.S. Geological Survey. U.S. Geological Survey Circular 1225. 1999. Washington, DC. 82 p.
7. Cohen, S.; Svrjcek, A.; Durborow, T.; Barnes, N. L. 1 *J. Environ. Qual.* **1999**, 28, 798-809.
8. Armbrust, K. L.; Peeler, H. B. *Pest Manag. Sci.* **2002**, 58, 702-706.
9. Branham, B. E.; Wehner, D. J. *Agron. J.* **1985**, 77, 101-104.
10. Cisar, J. L.; Snyder, G.H. *Crop Sci.* **1996**, 36, 1433- 1438.
11. Cole, J. T.; Baird, J. H.; Basta, N. T.; Huhnke, R. L.; Storm, D. E.; Johnson, G. V.; Payton, M.E.; Smolen, M. D.; Martin, D. L.; Cole, J. C. *J. Environ. Qual.* **1997**, 26, 1589-1598.
12. Evans, J. R.; Edwards, D. R.; Workman, S. R.; Williams, R. M. *Transactions of the ASAE.* **1998**, 41, 1323-1329.
13. Hong, S.; Smith, A. E. *J. Environ. Qual.* **1997**, 26, 379-386.
14. Moss, C. G.; Edwards, D. R.; Workman, S. R.; Williams, R. M. *Transactions of the ASAE.* **1999**, 42, 1631-1636.
15. Petrovic, A.M.; Young, R.G.; Ebel, J.C. Jr.; Lisk, D.J. *Chemosphere* **1993**, 26, 1549-1557.
16. Petrovic, A. M.; Guterunann, W. H.; Ebel, J. G. Jr.; Lisk, D. J. *Chemosphere* **1993**, 26, 1541-1547.
17. Smith, A. E.; Bridges, D. C. *Crop Sci.* 1996, 36, 1439-1445.
18. Watschke, T. L.; Mumma, R. O.; Linde, D. T.; Borger, J. A.; Harrison, S. A. In *Fate and Management of Turfgrass Chemicals*; Clark, J. M.; Kenna, M. P. Ed.; ACS Symposium Series 743; American Chemical Society: Washington, D.C., 2000, pp. 94-105.
19. Wauchope, R. D.; Williams, R. G.; Marti, L. R. *J. Environ. Qual.* **1990**, 19, 119-125.
20. Wu, L.; Liu, G.; Green, R. L.; Pacheco, P.; Gan, J.; Yates, S. R. *Pest Manag. Sci.* **2002**, 58, 335-342.
21. Petrovic, A.M.; Larsson-Kovach, I. *Chemosphere* **1996**, 33, 585-593.
22. Smith, M. C.; Bottcher, A. B.; Campbell, K. L.; Thomas, D. L. *Transactions of the ASAE* **1991**, 34, 837-847.
23. Duborow, T. E.; Barnes, N. L.; Cohen, S. Z.; Horst, G. L.; Smith, A. E. In *Fate and Management of Turfgrass Chemicals*; Clark, J. M.; Kenna, M. P.

Ed.; ACS Symposium Series 743; American Chemical Society, Washington, D.C., 2000, pp. 195-227.
24. Roy, J. W.; Hall, J. C.; Parkin, G. W.; Wagner-Riddle, C.; Clegg, B. S. *J. Environ. Qual.* **2001**, 30, 1360-1370.
25. Starrett, S. K.; Starrett, S. K. In *Fate and Management of Turfgrass Chemicals*; Clark, J. M.; Kenna, M. P. Ed.; ACS Symposium Series 743; American Chemical Society, Washington, D.C., 2000, pp. 180-194.
26. Carroll, M. J.; Hill, R. L.; Raturi, S.; Herner, A. E.; Pfeil, E. In *Fate and Management of Turfgrass Chemicals*; Clark, J. M.; Kenna, M. P.Ed.; ACS Symposium Series 743; American Chemical Society, Washington, D.C., 2000, pp. 228- 242.
27. Raturi, S.; Carroll, M. J.; Hill, R. L. *J. Environ. Qual.* **2003**, 32, 215-223.
28. Haith, D.A. *J. Environ. Qual.* **2001**, 30, 1033-1039.
29. Haith, D.A.; Andre, B. *J. Environ. Qual.* **2000**, 29, 1548-1554.
30. Haith, D.A.; Lee, P.C.; Clark, J.M.; Roy, G.R.; Imboden, M.J.; Walden, R.R. *J. Environ. Qual.* **2002**, 31, 724-729.
31. Walden, R.R.; Haith, D.A. *J. Environ. Qual.* **2003**, 32, 1138-1143.
32. Hargreaves, G. H.; Samani, Z. A. *Appl. Eng. Agric.* **1985**, 1, 96–99.
33. *The ASCE Standardized Reference Evapotranspiration Equation.* Allen, G.A.; Walter, I.A.; Elliott, R.; Howell, T.; Itenfisu, D. and Jensen, M. Eds. ASCE, Reston, VA, 2005.
34. Pagès, L.; Asseng, S.; Pellerin, S.; Diggle, A. In *Root Methods: A Handbook*; Smith, A.L.; Bengough, A.G.; Engles, C.; van Noordwijk, M.; Pellerin, S.; van de Geijn, S.C., Eds.; Springer-Verlag, Berlin. 2000, pp 113-135.
35. Carroll, M.J. Ph.D. Thesis, Cornell University. Ithaca, NY. 1989.
36. Prettyman, G.W. and McCoy, E.L. *Crop Sci.* **2003**, 43, 985-994.
37. Jury, W. A.; Focht, D. D.; Farmer, W. J. *J. Environ. Qual.* 1987, 16, 422-428.
38. Wauchope, R. D.; Rojas, K.W.; Ahuja, L.R.; Ma, Q.; Malone, R.W.; Ma, L.. *Pest Management Science* **2004**, 60, 222–239.
39. Saxton, K.E.; Washington State University. *Soil Water Characteristics*, version 6.1.52 2005. URL http://hydrolab.arsusda.gov/ soilwater/Index.htm
40. Saxton, K.E.; Rawls, W.J.; Romberger, J.S.;R. I. Papendick. *Soil Sci. Soc. Am. J.*, **1986**, 50, 1031-1036.
41. Moskal, T. D.; Leskiw, L.; Naeth, M. A.; Chanasyk, D. S. *Can. J. Soil Sci.* 2001, 81, 205–211.
42. *Smart Water Application Technology, Climatologically Based Controllers. 6th Draft Testing Protocol,* URL http://www.irrigation.org/gov/swat_display .aspx?sbjt=Controller&PD=6thDraftTestProtocols-ClimateBasedController .pdf
43. Green, R, L.; Gibeault, V.A. UCRTRAC Newsletter. June 1998, p 2.
44. Green, R.L.; Richie, W.E.; Gibeault, V.A. *Influence of Irrigation Scheduling on Bermudagrass and Zoysiagrass performance during the warm-season. Final Report.* URL http://ucrturf.ucr.edu/
45. Roy, J.W.; Parkin, G.W.; Wagner-Riddle, C. *Soil Sci. Soc. Am. J.* **2000**, 64, 86-93.

46. Agricultural Research Service. *Pesticide Properties Database.* 2000. URL http://www.arsusda.gov/acsl/services/ppdb/ppdb3.html
47. *The Pesticide Manual;* Tomlin, C., Ed.; British Crop Protection Council and The Royal Society of Chemistry, Cambridge, UK, 1994.
48. Hornsby, A.G.; Wauchope, R.D.; Herner, A.E. *Pesticide Properties in the Environment*; Springer-Verlag: New York, 1995.
49. *Herbicide Handbook, 7th ed.;* Ahrens, W.H., Ed.; Weed Science Society of America, Champaign, IL, 1994.
50. Horst, G.L.; Shea, P.J.; Christians, N.; Miller, D.R.; Stuefer-Powell, C.; Starrett, S.K. *Crop Sci.* **1996**, 36, 362-370.
51. Shannon, R.E. *Systems Simulation: The Art and Science.* Prentice-Hall, Englewood Cliffs, NJ, 1975, pp 224-227.
52. Nash, J.E.; Sutcliffe, J.V. *J. Hydrol.* **1970**, 10, 282-290.

Indexes

Author Index

Subject Index

D

M

N